Advances in Metal Matrix Composites: Structure, Properties and Applications

Advances in Metal Matrix Composites: Structure, Properties and Applications

Guest Editors

Lucia Lattanzi
Anders E. W. Jarfors

MDPI
Basel • Beijing • Wuhan • Barcelona • Belgrade • Novi Sad • Cluj • Manchester

Guest Editors
Lucia Lattanzi
Department of Materials and Manufacturing
Jönköping University
Jönköping
Sweden

Anders E. W. Jarfors
Department of Materials and Manufacturing
Jönköping University
Jönköping
Sweden

Editorial Office
MDPI AG
Grosspeteranlage 5
4052 Basel, Switzerland

This is a reprint of the Special Issue, published open access by the journal *Crystals* (ISSN 2073-4352), freely accessible at: https://www.mdpi.com/journal/crystals/special_issues/XVXRA49BMD.

For citation purposes, cite each article independently as indicated on the article page online and as indicated below:

Lastname, A.A.; Lastname, B.B. Article Title. *Journal Name* **Year**, *Volume Number*, Page Range.

ISBN 978-3-7258-7699-0 (Hbk)
ISBN 978-3-7258-7700-3 (PDF)
https://doi.org/10.3390/books978-3-7258-7700-3

Contents

About the Editors

Lucia Lattanzi

Lucia Lattanzi has been an Assistant Professor at the Department of Materials and Manufacturing, School of Engineering, Jönköping University since 2024. Lucia's research interests are the performance and production of Al-based alloys and their composites. Her work currently focuses on cast Al-Si alloys and their reinforcement with SiC particles to obtain metal matrix composites (MMCs), with research projects that also contribute to improving sustainability in the aluminum field.

Anders E. W. Jarfors

Anders E. W. Jarfors has been a Professor in foundry technology and casting at the Department of Materials and Manufacturing, School of Engineering, Jönköping University since 2012. His research areas encompass technology for the manufacturing of cast components, including both the process chain and the determination of their structure and properties, as well as phenomena that are crucial for their quality.

crystals

Editorial

Advances in Metal Matrix Composites: Structure, Properties and Applications

Lucia Lattanzi * and Anders E. W. Jarfors

Department of Materials and Manufacturing, School of Engineering, Jönköping University, Gjuterigatan 5, 55318 Jönköping, Sweden; anders.jarfors@ju.se
* Correspondence: lucia.lattanzi@ju.se

1. Introduction

Metal matrix composites (MMCs) are attractive materials due to their unique properties that stem from combining a wide range of matrix materials and reinforcements. The possibility of tailoring the mechanical response, thermophysical properties, chemical and electrical behavior, and wear properties makes MMCs incredibly interesting. The development of such materials has been driven by various applications in different fields, including automotive, aerospace, electrical and electronic applications. The challenge of creating a sustainable society lies in the recyclability of such materials.

Given the fast evolution of these materials and the numerous combination possibilities, the Special Issue "Advances in Metal Matrix Composites: Structure, Properties and Applications" [1] aimed to gather studies on the most recent advances in the field. Articles focusing on material modeling, microstructural characterization, wear testing, and mechanical and thermophysical performance at both room and elevated temperatures are welcome. Articles on the sustainability of MMCs, either by recycling the composite or using recycled matrices and reinforcements, are highly encouraged.

2. An Overview of Published Articles

An overview of the articles included in the present Special Issue is provided here, along with a brief description of the focus for each article. Among the review article and eight research articles, a variety of metal matrices and reinforcements are presented. Similarly, there is a wide variety of different applications for which each composite system is particularly suitable and of interest. Table 1 presents a summary of the main features investigated in the articles.

The review article by Chen et al. [2] focused on copper-based diamond composites, which have been the subject of numerous investigations for applications requiring higher thermal conductivity. However, the natural non-wetting behavior between diamond particles and copper matrix makes it challenging to fabricate copper-based diamond composites with high thermal conductivity. Thus, to promote wettability between copper and diamond particles, the copper/diamond interface must be modified by coating alloying elements on the diamond surface or by adding active alloying elements with carbon in the copper matrix. In this paper, we review the research progress on copper-based diamond composites, including theoretical models for calculating the thermal conductivity and the effect of process parameters on the thermal conductivity of copper-based diamond composites. The factors that affect interfacial thermal conductivity are emphatically analyzed in this review. Finally, the current problems of copper-based diamond composites and future research trends are recommended.

Table 1. List of the articles included in this Special Issue.

Article	MMC Matrix	MMC Reinforcement	Investigations
Chen et al. [2]	Copper (Cu)	Diamond particles	Literature review
Ghahremani et al. [3]	Aluminum-silicon (Al-Si)	Silicon oxide nanoparticles (SiO_2np)	Microstructure Wear
Golla et al. [4]	Aluminum-iron (Al-Fe)	Titanium carbide particles (TiCp)	Mechanical properties Wear
Du et al. [5]	Aluminum-silicon (Al-Si)	Silicon carbide particles (SiCp)	Microstructure Transfer efficiency
Hirsch et al. [6]	Aluminum-silicon (Al-Si)	Silicon carbide particles (SiCp)	Mechanical properties Heat treatment
Lewis et al. [7]	Aluminum (Al)	Silicon (Si)	Thermophysical properties
Yin et al. [8]	Magnesium-erbium (Mg-Er)	Nickel and zinc phases (Ni, Zn)	Microstructure Heat treatment
Zhuo et al. [9]	Tungsten-copper (W-Cu)	Tungsten fibers (Wf)	Mechanical properties

The research article by Ghahremani et al. [3] focuses on metal matrix nanocomposites (MMNCs), which are becoming materials of choice in various engineering and medical applications due to their superior combination of targeted properties. Among different MMNCs, aluminum-based composites are of particular importance. In many applications, relatively inferior wear limits the practical use of this valued metal. However, reinforcing aluminum and its alloys with ceramics, such as carbon allotropes, may largely circumvent these limitations. In the present study, an aluminum alloy A356/SiO_2 nanocomposite is fabricated using a vibration-assisted casting process, in which varying amounts of silica (SiO_2) nanoparticles, ranging from 0.125 to 0.375 wt.%, are added to the molten alloy. The use of ultrasonic power treatment had a significant influence on the microstructure, hardness, and wear properties. Microstructural and XRD analyses were performed on the fabricated monolithic and composite samples. To evaluate wear behavior, a hardness test and a pin-on-disk experiment were conducted on the samples under 60, 80, and 100 N forces at a constant speed of 1 m/s, with the sliding distance varying from 1000 to 2000 m. The abraded surfaces, wear debris, and EDS analysis were used to identify wear mechanisms. The samples with 0.125 wt.% exhibited the highest increase in hardness and the highest reductions in both friction coefficient and wear rate, by 52%, 50%, and 68%, respectively. The primary governing wear mechanism was abrasion, with limited evidence of delamination.

The research article by Golla et al. [4] highlights the development of high-performance materials in the construction materials industry, with a special focus on the production of aluminum matrix composites (AMCs) containing titanium carbide (TiC) particles. The stir casting method with ultrasonic assistance was employed to enhance the mechanical and tribological properties. ASTM standards were used to evaluate the influence of TiC particles on density, hardness (VHN), ultimate tensile strength (UTS), and wear resistance at various TiC weight fraction percentages (0.0 wt.%, 2.0 wt.%, 4.0 wt.%, 6.0 wt.%, and 8.0 wt.%). Field-emission scanning electron microscopy (FESEM) and X-ray diffraction (XRD) analysis were conducted to investigate the microstructural changes and elemental phases present in the synthesized composite. Results revealed that incorporating 8 wt.% TiC reinforcement into the metal matrix composites demonstrated significant improvements compared to the base alloy. A substantial enhancement in hardness by 32%, a notable increase of 68% in UTS, and a significant 80% rise in yield strength were observed when contrasted with the pure aluminum alloy. The tensile fracture analysis of the specimens revealed the presence of dimples, voids, and cracks, suggesting a brittle nature. To assess the wear characteristics of the composites, dry sliding wear experiments were performed using a pin-on-disk wear tester. Incorporation of TiC particles resulted in a lower coefficient of friction than the base alloy, with the lowest friction coefficient being recorded at 0.266 for 6 wt.% TiC, according to the data. FESEM and energy-dispersive X-ray spectroscopy (EDXS) were used to examine the surfaces of the worn pin. Overall, the inclusion of TiC reinforcement particles in the matrix alloy greatly enhanced the wear resistance and friction coefficient of the Al-6TiC composites. Plowing and adhesion occurred at lower loads, while delamination occurred at higher loads, as observed in the wear test.

The research article by Du et al. [5] examines lightweight SiC-particle-reinforced aluminum composites that have the potential to replace cast iron in brake disks, especially for electric vehicles. This study investigates the effect of SiC particle size and matrix alloy composition on the resulting transfer efficiency and particle distribution. The performance of a specially designed stirring head was studied using a water model, and the stirring head conditions were evaluated to understand the mechanisms of particle transfer and dispersion in molten aluminum. The standard practice of thermal pre-treatment promotes the wetting of the reinforcing particles and commonly causes clustering before the particles are added to the melt. This early clustering affects the transfer efficiency and particle dispersion, where their interaction with the melt's top-surface oxide skin plays a crucial role. Additionally, the transfer efficiency was linked to both the particle size and the chemical composition of the matrix alloy. Smaller particles aggravated the degree of clustering, and the addition of rare earth elements as alloying elements in the matrix alloy affected the particle dispersion. The stirring parameters should be selected to ensure cluster disruption when the carbides are added to the melt.

The research article by Hirsch et al. [6] investigates the solid-state sintering of cast aluminum powders using resistance heating sintering (RHS), also known as spark plasma sintering or the field-assisted sintering technique, which creates a very fine microstructure in the bulk material. This results in high-performance material properties, with improved strength and ductility compared to conventional production routes for the same alloys. In this study, the mechanical behavior of an RHS-sintered age-hardenable A357 (AlSi7Mg0.6) cast alloy and a SiCp/A357 aluminum matrix composite (AMC) was investigated. Aiming for high strength and good wear behavior in tribological applications, the AMC was reinforced with a high particle content (35 vol.%) of a coarse particle fraction ($d50 = 21$ μm). Afterwards, the separated and combined effects of particle reinforcement and heat treatment were studied under compressive load at both room temperature and 350 °C. At room temperature compression, the strengthening effect of precipitation hardening was about twice as high as that for the particle reinforcement, despite the high particle content. At elevated temperatures, the compressive deformation behavior was characterized by the simultaneous occurrence of temperature-activated recovery, recrystallisation, and precipitation processes. The occurrence and interaction of these processes were significantly affected by the initial material condition. Moreover, a rearrangement of the SiC reinforcement particles was detected after hot deformation. This rearrangement resulted in a homogenized dispersion of the reinforcement phase, with minimal particle fragmentation, offering potential for secondary thermo-mechanical processing of highly reinforced AMCs.

The research article by Lewis et al. [7], given the ever-growing emphasis on global decarbonization and rapid increases in the power densities of electronics equipment in recent years, studies the new methods and lightweight materials developed to manage heat load as well as interfacial stresses associated with coefficient of thermal expansion (CTE) mismatches between components. The Al–Si system offers an attractive combination of CTE performance and high thermal conductivity, while being a lightweight option. Such materials are of interest to industries where thermal management is a key design criterion, such as the aerospace, automotive, consumer electronics, defense, EV, and space sectors. This paper will describe the development and manufacture of a family of high-performance hypereutectic Al–Si alloys (AyontEX™) by a powder metallurgy method. These alloys are of particular interest for structural heat sink applications that require high reliability under thermal cycling (CTE of 17 μm/(m·°C)), as well as reflective optics and instrument assemblies that require good thermal and mechanical stability (CTE of 13 μm/(m·°C)). Critical performance relationships are presented, coupled with the microstructural, physical, and mechanical properties of these Al–Si alloys.

The research article by Yin et al. [8] investigates the influence of transition metals, such as nickel (Ni) and zinc (Zn), on the formation, morphology, and thermal stability of long-period stacking ordered (LPSO) phases in $Mg_{97}Er_2Ni_1$ and $Mg_{97}Er_2Zn_1$ alloys. In the as-cast state, both alloys consist of α-Mg and LPSO phases. Heat treatment at 540 °C for 20 h dissolves block-like and lamellar LPSO phases into the α-Mg matrix in the $Mg_{97}Er_2Zn_1$ alloy, with lamellar LPSO phases reprecipitating during subsequent cooling from 540 °C to 400 °C. A comparative analysis reveals that nickel significantly enhances the thermal stability of the LPSO phase compared to zinc. Nickel favors the formation of block-shaped LPSO phases, while zinc facilitates lamellar LPSO precipitation within the α-Mg matrix. The LPSO phase in the $Mg_{97}Er_2Ni_1$ alloy exhibits an exceptionally high melting temperature of 605 °C, the highest reported for an LPSO phase. Additionally, heat treatment at 500 °C for 100 h preserves the area fraction of the LPSO phase in the $Mg_{97}Er_2Ni_1$ alloy. At 540 °C for 100 h, the LPSO grains grow without phase dissolution or structural transformation of their 18R-type configuration. These findings provide valuable insights into the role of alloying transition metal elements in controlling the stability and morphology of LPSO phases, offering pathways for tailoring the morphology of the LPSO phase in the Mg-based alloys.

The research article by Zhuo et al. [9] investigates tungsten fiber (Wf) reinforced tungsten–copper (W-Cu) composites, which have broad application prospects in fields such as electronic packaging due to their excellent comprehensive properties. However, the correlation between the fiber parameters (content, aspect ratio, and orientation) and the mechanical behavior of the materials remains unclear. A combination of numerical simulation and experimental research was employed to construct a three-dimensional microstructural mechanism model and systematically investigate the influence of fiber parameters on the tensile properties and mechanisms of Wf/W-Cu composites. The results show that the critical fiber aspect ratio is 7.6. Below this value, the fiber pullout mechanism dominates, whereas above this value, fiber tensile fracture is the primary failure mechanism. As the fiber content increases from 1% to 6%, the tensile strength of the composite increases by 9.6%, and the yield strength increases by 10.2%, while the elongation after fracture decreases by 18.6%. As the fiber orientation angle increases from 0° to 90°, the material's strength initially increases and then decreases, whereas its toughness initially decreases and then increases. Short fibers achieve interface toughening through fiber pullout, crack deflection, and fiber bridging, while long fibers enhance the strength and toughness of the composite through load transfer and the fiber bridging effect. The damage evolution mechanism reveals the regulatory impact of fiber parameters on the multiscale mechanical behavior of the material. The research results can guide the optimization design of the composition and structure of Wf/W-Cu composites, providing new insights into the study of high-performance fiber composites and having significant implications for their engineering applications in extreme environments.

3. Summary

The current Special Issue "Advances in Metal Matrix Composites: Structure, Properties and Applications" collects research and review contributions about metal matrix composites. These materials are particularly attractive due to their unique and tailored properties from the combination of a wide range of matrix materials and reinforcements. The review article and eight research articles included in this Special Issue effectively reflect this variety of possibilities, exploring different combinations of metal matrices and reinforcements. Consequently, the articles focus on a wide variety of different applications and industrial sectors for which each composite system is particularly suitable and of interest.

Acknowledgments: As Guest Editors of the Special Issue (SI) "Advances in Metal Matrix Composites: Structure, Properties and Applications", L.L. and A.J. would like to express deep appreciation to all the authors who contributed valuable work to publish in the current SI, and all the anonymous reviewers who provided their professional opinions to support the peer evaluation process.

Conflicts of Interest: The authors declare that they have no conflicts of interest.

References

1. Crystals | Special Issue: Advances in Metal Matrix Composites: Structure, Properties and Applications. Available online: https://www.mdpi.com/journal/crystals/special_issues/XVXRA49BMD (accessed on 8 May 2025).
2. Chen, K.; Leng, X.; Zhao, R.; Kang, Y.; Chen, H. Progress in the Copper-Based Diamond Composites for Thermal Conductivity Applications. *Crystals* **2023**, *13*, 906. [CrossRef]
3. Ghahremani, A.; Abdullah, A.; Fallahi Arezoodar, A.; Gupta, M. An Investigation on the Enhanced Wear Behavior of Ultrasonically Stirred Cast A356/SiO_2np Nano-Composites. *Crystals* **2023**, *13*, 722. [CrossRef]
4. Golla, C.B.; Babar Pasha, M.; Rao, R.N.; Ismail, S.; Gupta, M. Influence of TiC Particles on Mechanical and Tribological Characteristics of Advanced Aluminium Matrix Composites Fabricated through Ultrasonic-Assisted Stir Casting. *Crystals* **2023**, *13*, 1360. [CrossRef]
5. Du, A.; Lattanzi, L.; Jarfors, A.E.W.; Zheng, J.; Wang, K.; Yu, G. On the Efficient Particle Dispersion and Transfer in the Fabrication of SiC-Particle-Reinforced Aluminum Matrix Composite. *Crystals* **2023**, *13*, 1621. [CrossRef]
6. Hirsch, S.J.; Berndt, N.; Grund, T.; Lampke, T. Combined Effect of Particle Reinforcement and T6 Heat Treatment on the Compressive Deformation Behavior of an A357 Aluminum Alloy at Room Temperature and at 350 °C. *Crystals* **2024**, *14*, 317. [CrossRef]
7. Lewis, P.; Tarrant, A.; Frehn, A.; Grensing, F.; Nicholson, J.; Farrah, N.; Acreman, M. Aluminium-Silicon Lightweight Thermal Management Alloys with Controlled Thermal Expansion. *Crystals* **2024**, *14*, 455. [CrossRef]
8. Yin, J.; Liu, Y.; Zhu, G.Z. Microstructural Evolution and Thermal Stability of Long Period Stacking Ordered Phases in $Mg_{97}Er_2Ni_1$ and $Mg_{97}Er_2Zn_1$ Alloys. *Crystals* **2024**, *14*, 1092. [CrossRef]
9. Zhuo, L.; Xie, Y.; Xu, H.; Luo, B.; Liu, N.; Chen, B.; Wang, H. Multi-Scale Numerical Simulation of Short Tungsten Fiber Reinforced Tungsten–Copper Composites: Influence Mechanisms of Fiber Parameters. *Crystals* **2025**, *15*, 209. [CrossRef]

crystals

MDPI

Article

An Investigation on the Enhanced Wear Behavior of Ultrasonically Stirred Cast A356/SiO_2np Nano-composites

Ahmad Ghahremani [1], Amir Abdullah [1,*], Alireza Fallahi Arezoodar [1,*] and Manoj Gupta [2,*]

1 Mechanical Engineering Department, Amirkabir University of Technology, Hafez Av., Tehran 15875-4413, Iran; a.qahremani@aut.ac.ir
2 Department of Mechanical Engineering, National University of Singapore, 9 Engineering Drive 1, Singapore 117576, Singapore
* Correspondence: amirah@aut.ac.ir (A.A.); a.fallahi@aut.ac.ir (A.F.A.); mpegm@nus.edu.sg (M.G.); Tel.: +98-21-64543419 (A.A.); +98-21-64543453 (A.F.A.); +65-6516-635 (M.G.)

Abstract: Metal matrix nanocomposites (MMNCs) are becoming the materials of choice in a variety of engineering and medical applications owing to their exhibiting a superior combination of targeted properties. Amongst different MMNCs, aluminum-based composites are of special importance. In many applications, a relatively inferior wear property limits the use of this valued metal in practice. However, reinforcing aluminum and its alloys by ceramics, carbon allotropes, etc., may circumvent these limitations to a great extent. In the present study, aluminum alloy A356/SiO_2 nanocomposite is fabricated by a vibration-assisted casting process, wherein varied amount of nanosilica, namely, 0.125, 0.25, and 0.375 wt.%, have been added to the melt. The use of power ultrasonic treatment had a great influence on the microstructure, hardness, and wear properties. Microstructural and XRD analyses were performed on the fabricated monolithic and composite samples. To evaluate wear behavior, a hardness test and pin-on-disk experiment were conducted on the samples under 60, 80, and 100 N forces at a constant speed of 1 m/s and the sliding distance was varied from 1000 to 2000 m. The abraded surfaces, wear debris, and EDS analysis were used to identify wear mechanisms. The samples having 0.125 wt.% exhibited the highest increase in hardness and the highest reduction in both friction coefficient and wear rate by 52%, 50%, and 68%, respectively. The main governing wear mechanism was abrasion, with limited evidence of delamination.

Keywords: ultrasonic treatment; wear; metal matrix composite; dispersion; reinforcement

1. Introduction

Metal matrix composites are some of the most suitable candidates for use in critical applications such as in automotive and aerospace industries. The growing need to lightweight materials for energy saving is a top priority of these industries [1,2]. Due to low density and specific strength, aluminum and magnesium are the main options for the base material in composite manufacturing [3,4].

Hard ceramics or carbon-based materials such as carbon nanotubes (CNTs) and graphite are used as reinforcement. The use of reinforcement mainly increases the strength and hardness values [5]. According to the literature, the size of the particles may also influence the mechanical properties. Compared to micron-size particles, nano-sized particles are preferred due to their unique properties and capabilities. If liquid state methods are used for processing, the particles may act as heterogeneous nucleation sites during solidification, thus refining the microstructure [6–8].

A variety of methods have been examined to produce the MMNCs, but the most common ones are casting and powder metallurgy. Casting is usually considered to be one of the most popular methods due to its low cost, high production capacity, simplicity and availability [9]. In addition to the mentioned features, there exist challenges to the production of cast metal matrix nanocomposite. The most challenging issue is the great tendency of the nanoparticles to agglomerate due to interparticle forces such as van der Waals and the lack of proper wetting between a given melt and reinforcing nanoparticles. Theoretically, a number of strategies have been presented to deal with the poor wettability, among which reducing surface tension, increasing the number of particles, reducing particle size, boosting solidification speed, and coating particles with materials having higher Hamaker constants can be mentioned [10]. In terms of wettability, the use of materials such as magnesium [11], titanium [12,13], calcium [14], etc., all increasing the surface energy of solid particles and reducing the surface tension, may improve wettability in the interfacial region of a solid–liquid composite system. Further, the melt temperature and wetting time are also effective parameters influencing the wetting behavior of particles by the liquid metal/alloy [15].

During processing, the chemical composition of material can be affected when wetting agents are added, and, accordingly, in some applications, in order to improve the wetting condition, one may not be allowed to use a reactive agent. Further, using reactive agents may produce unwanted compounds in the manufactured composite, adversely affecting the mechanical properties [14–16]. One of the other processes that is of great interest in increasing the dispersion quality of particles, as well as boosting the affinity between a given melt and the reinforcing solid particle, is mechanical stirring [17]; in particular, high-power ultrasonic waves are used to stir the composite slurry [18,19]. In vibration-assisted casting, a stirrer with a material resistant against melt attack or a metallic stirrer coated by a resistant ceramic material is used. The main parameters of mechanical stirring are the geometry and size of the blades, the depth where the stirrer is placed, and the stirring speed and time [16,20–26]. For example, Karbalaei et al. [17] investigated the effect of stirring time on the wear resistance of stir cast A356/Al_2O_3 composite and found that long stirring time is useful in terms of applied shear forces; however, with respect to gas absorption and thus to increasing porosities, longer stirring may have a negative effect. They concluded that stirring has an optimal time value. The use of high-power ultrasonic waves is considered as a mechanical stirring process, wherein cavitation is responsible for deagglomerating the particles and dispersing them uniformly throughout the host melt matrix, bringing opportunities for heterogeneous nucleation [18,27–29]. In addition to the deagglomeration and uniform distribution of particles in the melt [29,30], other phenomena such as degassing [31], increasing the surface energy of solid particles [18], cleaning surfaces from contaminants [32,33], preventing segregation [34], the reduction of thermal stresses [35], the uniform distribution of different phases in the bulk [36] and the creation of a strong acoustic flow in the melt, mixing melt slurries [37–39], have been mentioned in the literature. Mechanical methods, especially the effective use of ultrasonic waves, are more efficient and less problematic compared to other methods, especially for realizing a fine and uniform distribution of nanoparticles in a given cast MMNCs [40].

In this study, A356 alloy has been chosen due to its high castability [41]. Moreover, as mentioned earlier, adding elements such as Mg, Ti, Ca, and Zr causes better wettability in metal–ceramic nanocomposites [42,43], and these elements are mainly present in A356 composition. For strengthening, nanosilica is used as reinforcement, being relatively inexpensive and available, and, as far as we know, less research has been conducted on it. Compared to SiC, nanosilica has a lower contact angle with molten aluminum, as indicated by Tekmen et al. and Hashim et al., who tried to oxidize the SiC surface by thermal treatment to create a better wettable SiO_2 layer on the silicon carbide surface [32,42]. Ultrasonic-assisted casting is used in this work to uniformly distribute silica nanoparticles in A356 melt. Since little is known of the mechanical and tribological behavior of the A356/SiO_2 nanocomposite, the present paper primary addresses the wear response of the ultrasonically stirred and cast A356/SiO_2 composites.

2. Materials and Methods

A356 aluminum alloy is used as the base matrix alloy, being supplied from the Materials Engineering Department at Isfahan University of Technology, Isfahan, Iran. The chemical composition of the alloy is given in Table 1. In each casting experiment, 1600 g of A356 alloy was poured into a graphite crucible to be melted under the protection of argon gas with a purity of 99.99%. Upon raising the melting temperature to 750 °C, 98.5% purity silica nanoparticles, supplied by Fadak-Isfahan New Technologies Complex, with the specifications presented in Table 2, were added to the melt (0.0, 2.0, 4.0, and 6.0 g). The corresponding amounts of the added reinforcing agent are, respectively, 0.0%, 0.125%, 0.25%, and 0.375%. The mentioned samples are named, respectively, AMNC0, AMNC2, AMNC4, and AMNC6 for ease of review. Figure 1 shows TEM and SEM images of the used silica nanoparticles.

Table 1. Chemical composition of A356 aluminum alloy.

Zn	Mn	Cu	Fe	Ti	Mg	Si	Al	Material
0.03	0.05	0.1	0.15	0.2	0.3	7.5	Balance	wt.%

Table 2. Specifications of nanosilica reinforcement.

Value	Unit	Properties and Test Method
210–240	m^2/g	Specific surface area (CTAB adsorption)
30–50	nm	Main particle size (TEM)
2.65	g/cm^3	Density
0.13–0.15	g/cm^3	Tamped density ISO-787-11
1.3	Wm^{-1} K	Thermal conductivity
12.3	10^{-6} K^{-1}	Thermal expansion coeff.
1830	°C	Melting point
≤98.5	%	SiO_2 Content ISO 3262/17

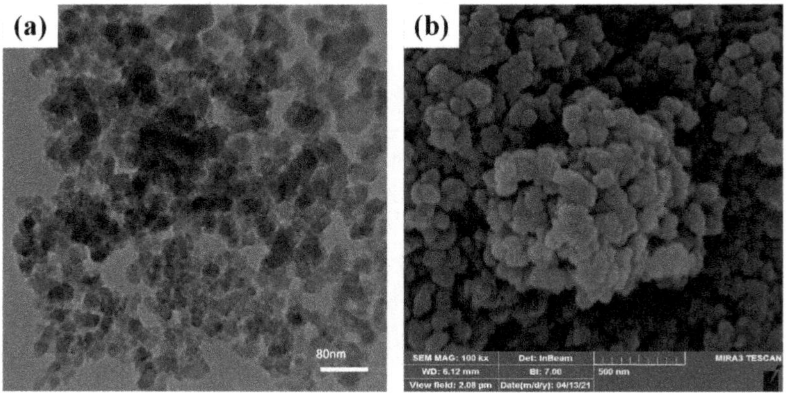

Figure 1. (**a**) TEM and (**b**) SEM of the nanosilica reinforcement used in this study.

The A356/SiO_2 composite samples were fabricated by ultrasonic-assisted casting, wherein a high power 2.0 kW piezoelectric transducer with a resonant frequency of 20 kHz was used. The transducer was first designed along with its vibration horn. After simulating and determining the exact geometry, the horn concentrating the ultrasonic vibrations was fabricated. There were six ring piezoelectrics that were tightened by a steel bolt between the steel backing part and the titanium matching part. The ultrasonic horn made of a Ti6Al4V alloy with specific geometry and dimensions was connected to the transducer in part of the vibrating node (see Figure 2a). The transducer with such configurations can also be used in some other manufacturing processes, including powder consolidations, machining, welding, etc. [44–47]. To protect the transducer, it was placed inside metal housing and the concentrator was placed outside the mentioned housing. To prevent piezoelectric damage from the generated heat, as well as from the heat transferred from the melt, the cooling system on the chamber is designed and installed exactly at the location of the piezoelectrics. In this system, cooling air is blown into the chamber and a fan was installed at the end of chamber to exhaust the air. The vibration head of the concentrator was immersed in the melt at a temperature of 750 °C and sonication was performed for 5.0 min. It should be noted that this period of time was optimized after trial and error. In other words, for times less than 5.0 min, the operation of ultrasonic waves was not effective enough, and for longer times, the probability of Ti dissolving at the end of the horn increased at high temperatures and high frequency vibration, causing contamination of the melt and unwanted effects on the microstructure. In addition to the issue of the melt being contaminated by Ti, the possibility of piezoelectrics being damaged under long-time exposure increases and can cause the operation to stop.

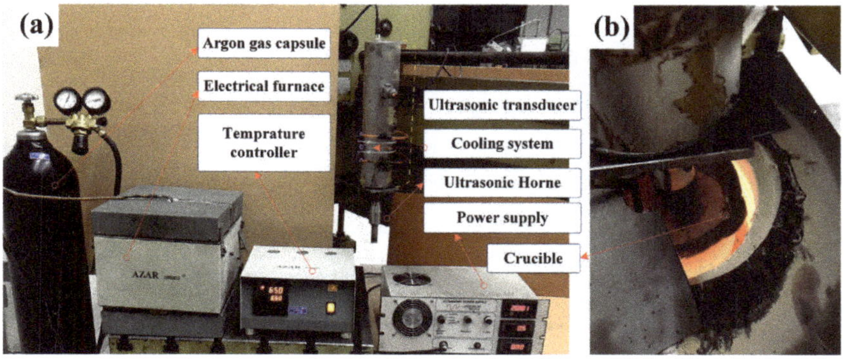

Figure 2. (**a**) Experimental setup and (**b**) A356/SiO_2 melt under sonication.

The layout of the experiment can be seen in Figure 2. A356 aluminum ingots were first placed in the graphite crucible, after accurate weighing, to be melted by an electric furnace under the protection of argon gas. Silica nanoparticles were then preheated (at 250 °C for 1 h), after accurate weighing by a scale with a resolution of 0.1 mg, so that the moisture was removed; finally, they were gradually added to the melt at the same time as applying ultrasonic vibrations at a temperature of 750 °C. After the melt was treated by ultrasonic vibrations, the composite slurry was immediately poured into a steel mold with an internal cross section of 60 × 60 mm, a wall thickness of 10 mm, and a height of 200 mm. After leaving the cast materials from the mold, the cooled ingots were machined and cut by a wire-cut machine (Wire-EDM, Tabriz Machines Co., Tabriz, Iran). They were then used to perform hardness measurements, wear tests, and microstructural analyses.

Characterization

Vickers hardness tester (DVKH-1, OGAWA SEIKI Co., LTD., Shinjuku-ku, Japan) with 5 kg weight and 15 s dual time was used in accordance with ASTM E92-17. To ensure repeatability, the tests were conducted with 5 repetitions per sample. Moreover, the hardness curve was drawn in the longitudinal direction of the cast ingots. To estimate the porosities of the cast samples, Archimedes test was performed.

A standard pin-on-disk apparatus, supplied by Advanced Modern Industry Company, Tehran, Iran, was used to measure the wear rates of the samples. The schematic and real images of the apparatus are shown in Figure 3. One of the special features of the mentioned device is the placement of the disc vertically; hence, almost all wear debris may not remain on the surface of the disc due to gravity and centrifugal force. In fact, in those wear experimental setups with a horizontal disc, some worn particles and hard reinforcing particles are separated but remain on the surface of the disc, and they may enter the space between the disc and the pin, leading to the scratching of the pin surface or providing conditions for three-body abrasive wear, capable of inducing errors either in the wear results or in wear mechanisms [48]. In this device, power is supplied through a linear actuator and the values of 10 to 100 newtons can be applied. The friction force can also be measured through a load cell that is parallel to the disk surface. A Pt100 type sensor is also installed to measure the temperature.

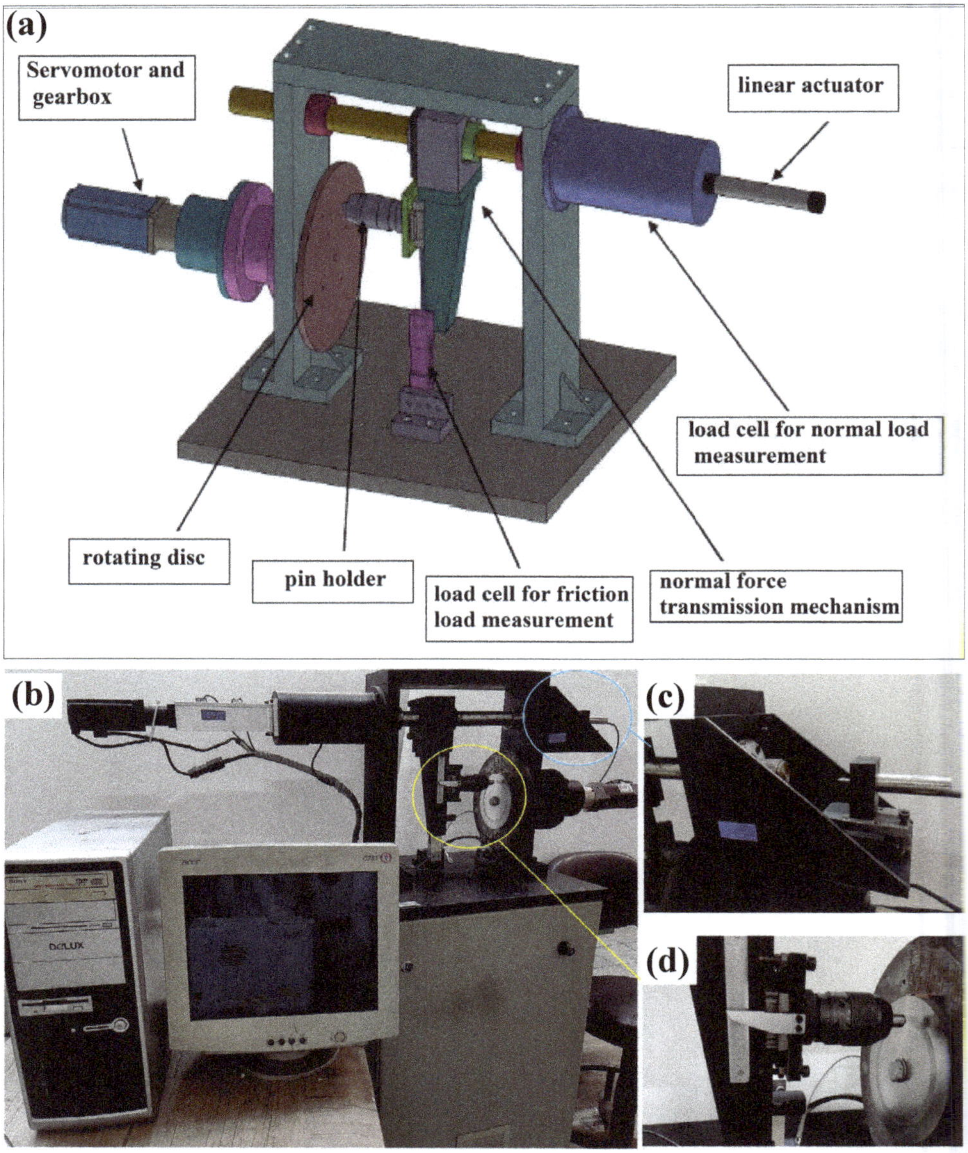

Figure 3. (**a**) Schematic of device with pin on a vertical disc, (**b**) actual image of the device, (**c**) wear measurement sensor, and (**d**) pin and disc in higher magnification.

The samples, in the form of pins, were prepared and gripped by a special holder. The abrasive disc was made of AISI 52,100 steel with the hardness of 65 HRC, diameter of 180 mm, thickness of 15 mm, and the surface roughness of 3 μ. This disk rotates at a maximum speed of 300 rpm, having the ability of closed loop control of an accuracy of 0.01%. The disc is perfectly perpendicular to the pin. A sensor is installed in this device to show the wear volume based on the amount of wear of cylindrical samples with micron accuracy. The samples were prepared in the form of pins with dimensions of 8 mm in diameter and 25 mm in height. To obtain accurate results, the end of each pin was spherically machined and polished with SiC sandpapers from 240 to 3000 grades (see Figure 4).

Figure 4. Fabricated samples: (**a**) before wear test; (**b**) after wear test.

The samples were first cleaned with a soft cloth just before and after the wear tests. They were then washed with acetone to remove contaminants and dried using cold air. Finally, they were weighed with a resolution of 0.1 mg. The weight difference was analyzed and studied as the main indication of wear values. The wear test parameters are given in Table 3.

Table 3. The parameters used in wear experiments.

Distance (m)	Speed (m/s)	Normal Force (N)	Nanosilica (g)	Parameter
1000, 1500, 2000	1	60, 80, 100	0, 2, 4, 6	Value

In order to analyze the microstructure of the cast samples, metallographic photographs were taken by an Olympus PME-3 optical microscope (LECO Co., Michigan, MI, USA) at a magnification of 100×. SEM images were also taken from the primary samples to check the microstructure for abraded surfaces and to estimate the wear mechanisms by FE-SEM MIRA3, TESCAN Co. (Brno, Czech Republic). In order to further investigate the effect of silica reinforcements, XRD analysis was performed using Inel, EQUINOX 3000, France, working with a voltage of about 40 kV and a current of 30 mA. The wavelength and the range of the mentioned XRD device are, respectively, λ = 0.154 nm and 2θ = 0–110°, with a step size of 0.01°.

3. Results and Discussion

3.1. Microstructural Evaluations

After machining, the cast ingots were cut by a wire-EDM to prepare the standard samples with desired dimensions. These samples were ground using 220- to 3000-grade emery clothes and then polished with a polishing machine using 0.3-micron alumina suspension. Upon washing and drying, the specimens were etched by a Keller's agent.

Figure 5 shows the microstructure with 100× magnification of AMNC0, AMNC2, AMNC4, and AMNC6.

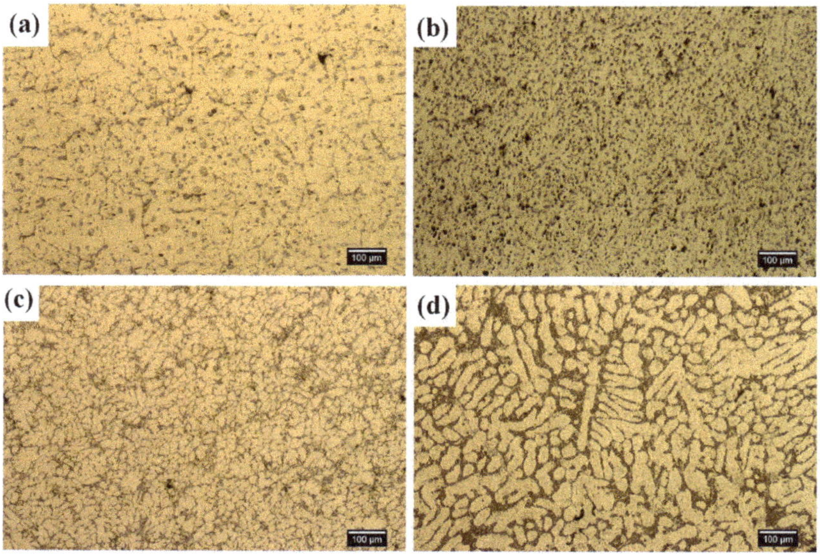

Figure 5. Optical microscopy results: (**a**) AMNC0, (**b**) AMNC2, (**c**) AMNC4, and (**d**) AMNC6.

The microstructures given in Figure 5 show that the microstructure of the A356 aluminum composite could substantially change only if 0.125 wt.% silica nanomaterial was added to the melt. Reduction of the grain size and break down of the dendritic structure is observed in this weight fraction, owing to the application of powerful and effective ultrasonic irradiation in deagglomerating and dispersing the silica particles evenly throughout the melt slurry. The uniform dispersion and distribution of nanoparticles may considerably facilitate heterogeneous nucleation, resulting in a finer microstructure. Cavitation being one of the main consequences of ultrasonic treatment may break the dendritic arms and structures. By adding more nanoparticles to the melt in the AMNC4 nanocomposite, no improvement was seen in the microstructure (Figure 5c); a coarser structure and the dendritic structure appeared. In fact, by adding nanoparticles twice, the viscosity of the melt increased, and as a result, the ability of ultrasonic dispersion decreased slightly, and the structure become coarser to some extent. For the same reason, by adding more nanoparticles, more dendritic and coarser structures are seen, owing to a decrease of ultrasonic efficiency in more viscous composite slurry (Figure 5d). With this, it can be seen that the optimal percentage of nanoparticles in these conditions is 0.125 wt.%.

3.2. XRD Results

A comparison of the XRD patterns of the different samples is depicted in Figure 6. Since the percentage of the added nanoparticles was very low, and since the nanoparticles were very small and fine, no peak related to silica was observed in any of the XRD patterns. Hosseini and his colleagues have also reported the absence of peaks related to the reinforcing phase, attributing it to the small size of the nanoparticles and the lower weight fraction of the reinforcing particles [49].

The greater broadening and lower intensity of the peak of the 0.125 wt.% confirm that the greatest reduction was achieved due to the addition of nanoparticles in the AMNC2 sample. The addition of reinforcements may cause the distortion in the structure of the nanocomposite, and the result is a broadening of the XRD pattern of the nanocomposite compared to the monolithic alloy [50]. Based on an XRD analysis, Sharma et al. [51] showed that the pattern of XRD peaks in the Al6082/SiC nanocomposite shift a little towards

2θ angles—lower than in the un-reinforced alloy. Similarly, the shifts towards lower angles (left side) are seen when SiO_2 is added. This pattern has also been reported by others in the literature [50].

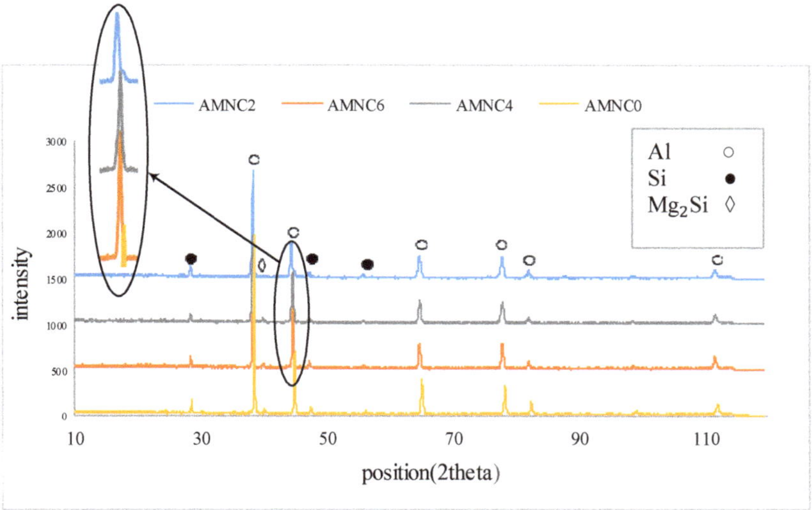

Figure 6. X-ray diffraction patterns of the A356 alloy and Al356/SiO_2 nanocomposites with 0.125, 0.25, and 0.375 wt.% nanosilica.

3.3. *Hardness Results*

To determine the hardness changes along the length of the cast sample, a hardness test was performed at each 2 cm interval. After sectioning the surface into three separate zones, i.e., the perimeter, middle points, and the center of the piece, the hardness test was performed, and the test was repeated at least five times in each zone. The results are used to obtain the hardness profile along the length of the samples. Figure 7 shows the hardness profile of each monolithic and composite sample.

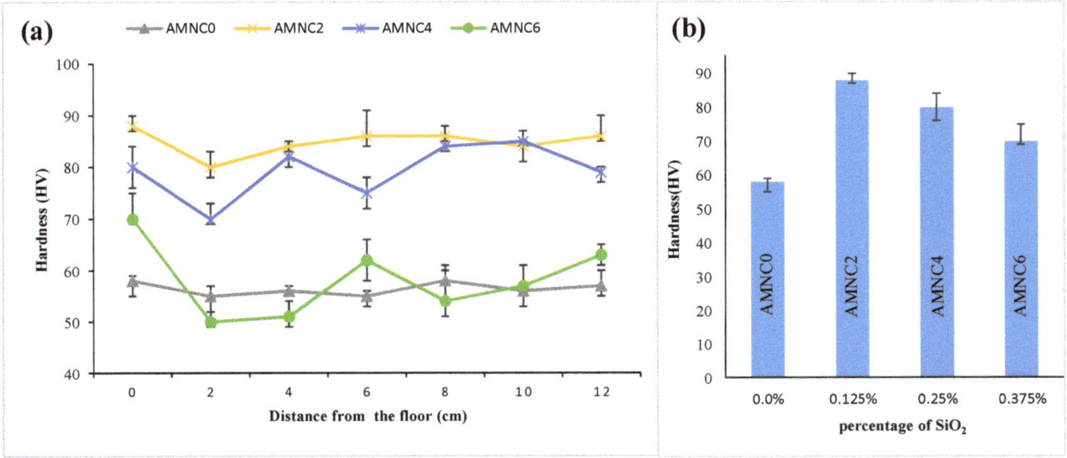

Figure 7. (**a**) Hardness profile along the length of the cast ingot; (**b**) comparison of samples AMNC0, AMNC2, AMNC4, and AMNC6 prepared from the bottom of the cast ingot.

As seen in Figure 7, the composites with different weight fractions showed higher hardness values than the base metal. The highest hardness value of 88 HV was seen in the AMNC2 nanocomposite, but the value decreased with an increase in silica content, reaching to the lowest amount of 68 HV in the AMNC6 sample. The microstructure shown in Figure 5 also confirms that the structure of the AMNC2 sample is quite fine. The fine structure based on the Hall–Patch relationship contributes to the greater strength of the nanocomposite [52]. On the other hand, the added particles can increase the strength of the nanocomposite through other mechanisms such as Orowan, CTE (coefficient of thermal expansion), and EM (elastic modulus) [35]. With an increase in the wt.% of reinforcement, the viscosity of the aluminum melt increases, and as a result, ultrasonic waves become less effective than when the viscosity of the fluid is higher [27]. Moreover, with the addition of nanoparticles, porosity may have increased; therefore, it can be concluded that gradually, with the addition of nanosilica, the degree of nanosilica dispersion may tend to decrease and the severity of porosity increases. Others have also reported that by adding relatively small percentages of ceramic nanoparticles, the hardness shows a significant increase, and with further addition of reinforcing phase, the increasing trend slowed down or had a decreasing trend [52–54].

The noteworthy point in the hardness profile of Figure 7a is that the hardness is higher ate the bottom and top of the cast samples, and there are fluctuations along the length of the sample, which mainly depend on the cooling conditions and the weight fraction of nanoparticles in that area. The reason for the higher hardness in the bottom can be due to the following reasons: (i) The bottom is the first place where the melt is in contact with the mold; hence, the cooling speed in the bottom is higher than any other points; it means faster cooling leads to finer structure and thus higher hardness of the samples. In this project, the steal mold is cooled by water from its bottom part, which contributes to the higher temperature gradient in this part. (ii) One of the mechanisms of engulfment of small particles in the molten metal, based on the capture theory, is the viscous capture mechanism, stating that if the cooling rate is higher than a critical rate, the possibility of particle engulfment at the solidification front will be higher [10]. Considering the mentioned theory, more particles were probably engulfed in this area and harder nanocomposites formed in this area. In the case of greater hardness in the upper parts, it is also possible to raise the possibility that the solidification front is from the lower part of the mold, that the nanoparticles are possibly rejected as impurities, and that the upper part will be richer from nanoparticles than the middle parts, and the observed difference in hardness is due to the higher amount of nanoparticles at the top of the solidified sample.

The Archimedes test was used to measure the density of the cast samples. The test is based on weighing a sample in two different fluids such as air and distilled water. Generally, the reference fluid is air, and the second fluid is distilled water. Each test was repeated three times; each sample was carefully washed and dried with washing liquid and acetone before testing. We obtain the density of the sample (ρ_s) based on equation 1, wherein m_{air} is the weight of the sample in air, m_{dw} is the weight of the sample in distilled water in the immersion state, and ρ_{dw} is the density of distilled water. The air density is neglected.

$$\rho_s = \rho_{dw} \times \frac{m_{air}}{(m_{air} - m_{dw})} \quad (1)$$

The Archimedes test was performed for the monolithic sample and nanocomposites with different nanosilica weight fractions; the results are shown in Table 4.

Table 4. Porosity percentage of monolithic and nanocomposite samples, i.e., AMNC2, AMNC4, and AMNC6.

AMNC6	AMNC4	AMNC2	AMNC0	Samples
2.5	1.6	1.2	0.6	Porosity(%)

3.4. Wear Response

Researchers have reported that the use of hard ceramic reinforcements—especially those in nanometric scales—in aluminum matrix nanocomposites may protect the soft matrix phase against wear, reducing the severe surface shear strains typically observed in the unreinforced samples. Hard ceramic particles may act as an abrasive member and load bearer. The microstructure of the base alloy and its history, such as the manufacturing route, as well as mechanical and heat treatments, may have a great influence on the wear resistance of its composite material. This effect is mainly through the microstructural change, the dispersion of reinforcing particles, voids and porosities, the bonding in reinforcement–matrix interfacial region, and the mechanical properties [5,55].

A common criterion, i.e., weight loss, was used to compare and study of the wear of samples. Figure 8 demonstrates the changes in the weight loss of the samples under different conditions.

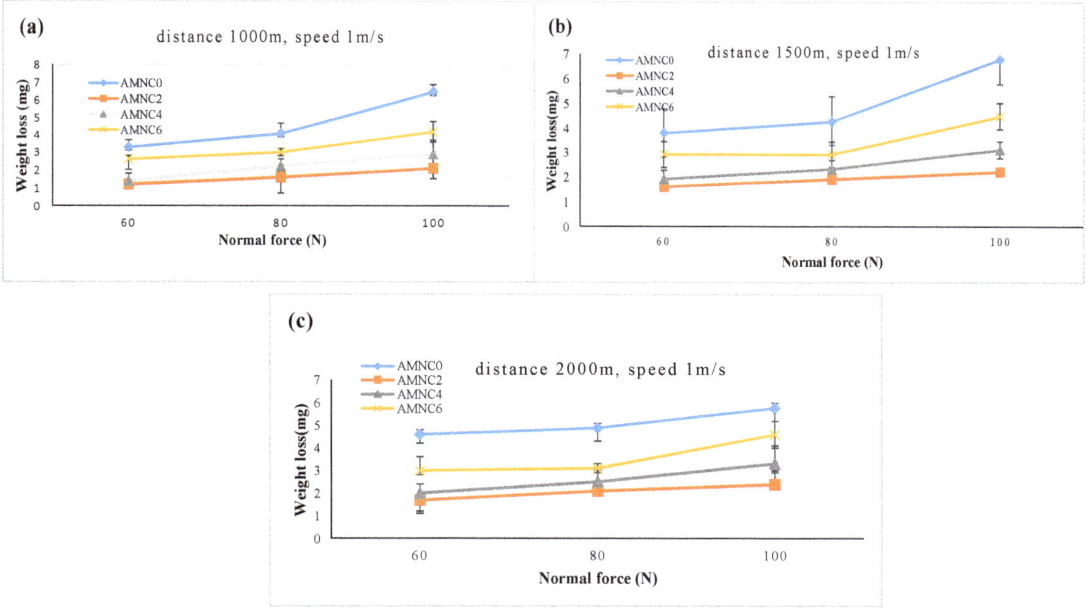

Figure 8. Weight loss comparison versus the vertical force of wear at a speed of 1 m/s and the distance of (**a**) 1000 m, (**b**) 1500 m, and (**c**) 2000 m.

To obtain the wear rates and for a better comparison between the composite and monolithic samples, Equation (2) is used.

$$Wear\ rate = \frac{Weight\ loss \times 10^{-3}}{\rho \cdot D} \quad (2)$$

In Equation (2), wear rate is in mm^3/m, and weight loss is in g; ρ is the actual density of a given sample being measured by the Archimedes test in g/cm^3, and D is the distance traveled in meters.

The wear rate in terms of force in different distances of 1000, 1500, and 2000 m is presented in Figure 9.

The wear rate can change according to Archard relationship and the hardness of the samples [56–58]. According to the hardness result, the addition of up to 0.125 wt.% nanosilica may considerably increase the hardness, but it tends to gradually decrease when further reinforcement is added to the aluminum melt. The wear rate also follows the same hardness trend. There is a significant difference between the wear rate of the AMNC2 sample and that of the A356 monolithic alloy. For example, the wear rate at a distance of 1500 m and a force of 100 N has decreased from about 0.00168 mm^3/m in the unreinforced alloy to 0.00054 mm^3/m in the AMNC2 nanocomposite, attaining more than three times the reduction in the wear rate. Despite the fact that the porosity of AMNC2 is slightly higher than the monolithic material, and that the porosity usually has a negative effect on the wear resistance, the extreme reinforcement-induced hardness and microstructural refinement in the AMNC2 sample may finally lead to a significant decrease in the wear rate. Due to lower viscosity, it seems that ultrasonic irradiation may treat more effectively the base alloy melt or the nanocomposite melts with lower silica weight fractions. The ultrasonic treatment is able to create a more homogeneous structure through the cavitation mechanism, wherein the structure can be considerably refined, and those clustering and agglomerations can be effectively disintegrated and finely dispersed in the matrix. In addition to deagglomeration, ultrasonic waves can also cause degassing to some extent. The phenomenon of ultrasonic degassing causes stronger bonding between the nanoparticles and the matrix, leading to better load transfer from matrix to nanoparticles [18]. Adding nanoparticles may cause micro-void nucleation around the nanoparticles. If nanoparticles are agglomerated, there might be more porosity inside the aggregates; consequently, they prevent the flow of melt and cause the formation of cavities [17]. By applying vibrations with a frequency of 20 kHz, ultrasonic waves can mitigate the porosity originating around the nanoparticles.

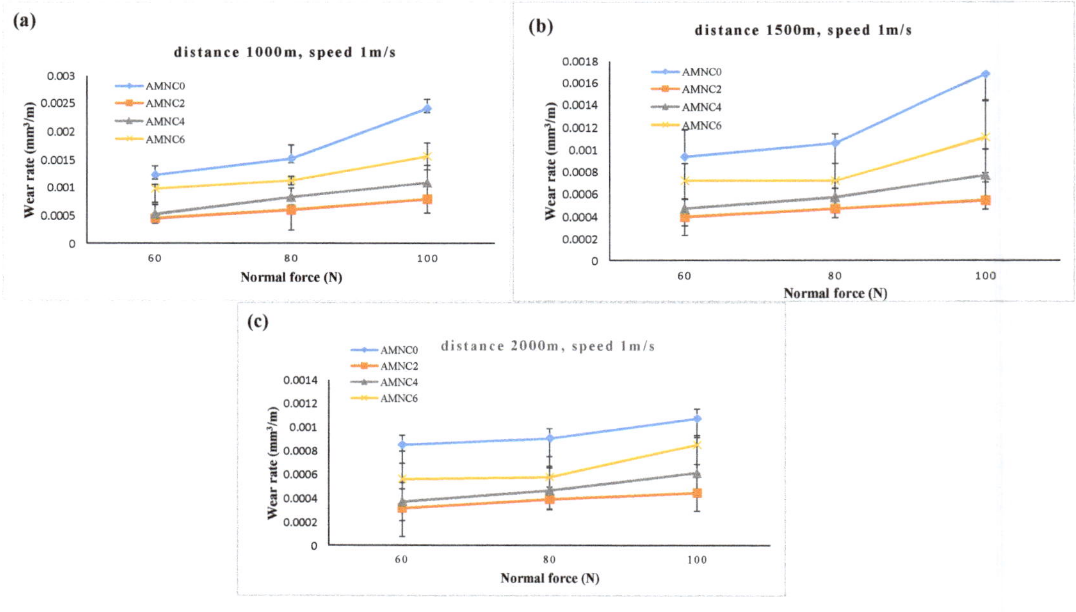

Figure 9. Wear rate versus the normal force at the speed of 1 m/s and a distance of (**a**) 1000 m, (**b**) 1500 m, and (**c**) 2000 m.

An increase in the normal force leads the plowing force (or friction force) to increase. In these conditions, more penetration into the material occurs and sub-surface fracturing will be expected; further, delamination may be activated when more shear strain is applied [59]. It is worth noting that the wear resistance improvement is higher at greater forces. In other words, nanoparticles have a greater effect on the load bearing capacity at higher wear forces and exhibit more wear resistance [59,60]. For example, at a distance of 1500 m and a force of 60 N, a 57% reduction in the wear rate is seen in AMNC2 compared to the unreinforced alloy, while the wear reduction rate under the same conditions and a force of 100 N can be up to 68%.

3.5. Coefficient of Friction

Figure 10 shows the coefficient of friction (COF) for the unreinforced alloy and the nanocomposites of AMNC2, AMNC4, AMNC6 in terms of distance 1000 m and under 60 N force. In the first stages, because the surfaces are not yet fully involved, lower values of the friction coefficient are observed, but then it increases quickly, and after this stage, a large discrepancy is seen in the values. The average values are depicted by the non-continuous line. The average value of the coefficient of friction in terms of the nanosilica weight fraction is shown in Figure 11a. An important point is that all the composite samples have a lower friction coefficient than the unreinforced sample. The COF of the monolithic sample is about 0.52, and it reached 0.27 in the AMNC2 nanocomposite, attaining a 50% reduction. The non-continuous line that indicates the average amount of COF of the AMNC2 sample is increasing. It is possible that with increasing distance and temperature due to the low weight fraction, the resistance to tangential forces has decreased and the friction coefficient has increased slightly. By increasing the weight fraction in AMNC4 and AMNC6 samples, the friction coefficient increased slightly. The COF itself is affected by the sum of the factors that are given in a simple form of the following relationship [61].

$$\mu = \mu_{el} + \mu_{pl} + \mu_{ad} + \mu_d \quad (3)$$

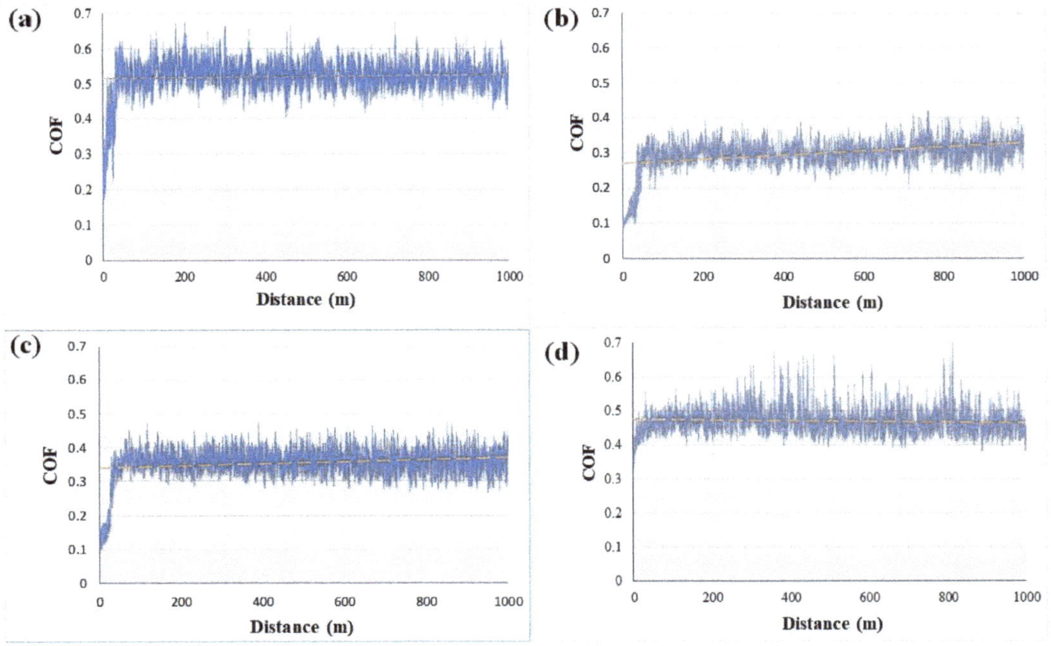

Figure 10. COF versus by the distance for: (**a**) AMNC0, (**b**) AMNC, (**c**) AMNC4, and (**d**) AMNC6. The average values of COF depicted by dotted line.

In this regard, μ_{el}, μ_{pl}, μ_{ad}, and μ_d are, respectively, related to elastic deformation, plastic deformation, adhesion between Al particles, and the effect of silica nanoparticles. As is clear from Figure 5, the AMNC2 nanocomposite has a finer structure and according to the Hall–Patch equation; it can be inferred that the elastic deformation term becomes greater. It has also been observed that the mechanical strength of nanocomposites has increased compared to the base alloy, as is confirmed by the hardness test results. Therefore, the term of plastic deformation is also higher in composite samples. In the meantime, increasing the hardness and toughness of the material leads to a lower contact in the mating surfaces; consequently, the adhesive forces attracting the two surfaces decrease, and it can be concluded that the value of μ_{ad} may also reduce in the composite samples [62]. Finally, according to references [63–65], the presence of SiO_2 particles has a lubrication property, reducing the friction coefficient between the two surfaces. Considering why the COF does not decrease with the increase in the nanosilica weight fraction from 0.125 to 0.375, it can be noted that despite the presence of more nanoparticles, the porosity also increases, and the reason is the lower efficiency of ultrasonic waves at higher viscosities. Probably, a higher porosity reduces interfacial bonding, and the mentioned clustered nanoparticles/agglomerates are separated from the matrix due to the tangential forces, and, in a way, it provides three-body wear conditions between the two surfaces. As a result, scratching the surface by the mentioned large agglomerates, the surface roughness may increase in higher weight fraction composites [17]. By increasing the nanosilica weight fraction, in addition to increasing the porosity rate, there is also the possibility of particle agglomeration, being separated from each other due to lower shear forces, and this may cause more surface destruction [66]. Therefore, in summary, we can point to factors such as increasing the porosity and agglomeration in the weight fractions higher than the sample with the optimum silica reinforcement (here 0.125% by weight), which do not allow further wear resistance improvement.

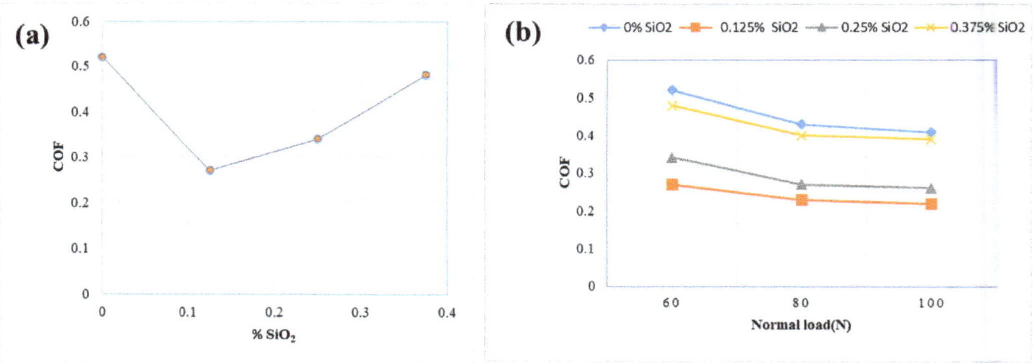

Figure 11. (**a**) COF versus nanosilica weight fraction; (**b**) friction coefficient versus normal force.

COF has high scattering values in all the samples, probably due to the sticking and separation of pin and disc surfaces. The oscillation range in the AMNC2 composite sample is much narrower compared to the monolithic alloy, indicating much more uniformity in the mentioned weight fraction.

Figure 11b shows the variation of COFs in terms of normal forces. It is seen that an increase in normal force results in a decrease in COF. The slope of the reduction of the COF is greater from the force of 60 to 80 N. The change of the COF in the AMNC2 sample does not considerably change, while the unreinforced sample has decreased significantly due to the wear force. Naturally, when the normal force increases, the tangential force, the elastic and plastic deformation, and the penetration depth of the asperities of the two surfaces may increase accordingly, being responsible for the rising temperatures at the contact point. As the temperature goes up, local softening occurs and the tangential force may reduce [67,68]. The reduction of the COF induced by the increased normal force has also been observed by others in the literature [50,69,70]. Karbalaei Akbari [53] has reported the mentioned phenomenon while studying the tribology of Al356/TiO_2 and TiB_2 nanocomposites, attributing it to the formation of a tribo-oxide layer as a result of the increasing force.

3.6. Wear Mechanisms

The SEM results of the abraded surfaces under the speed of 1 m/s, the distance of 1500 m, and the force of 80 N are shown in Figure 12. The mechanisms observed in these samples are a combination of the adhesive mechanism, delamination, and abrasion. The wear path can be recognized by an arrow. In the lower left corner of each figure is a lower magnification image demonstrating the amount of material removed by the wear test. The surface of the unreinforced alloy shows a significant degradation and there is almost no trace of the scratch mechanism. The creation of deep pits and plastic deformation, as well as delamination with large thicknesses, being the characteristics of the adhesive and delamination mechanisms, respectively, can be clearly detected. Looking at the image with lower magnification in the left and bottom corner of Figure 12a, it can be seen that the flash is created along the wear path which somehow refers to the plastic deformation of the base alloy and stretching in the wear path. When metal-to-metal contact occurs, material transfer occurs between the two surfaces in contact, and the dominant mechanism in this type of contact is mainly adhesive [71].

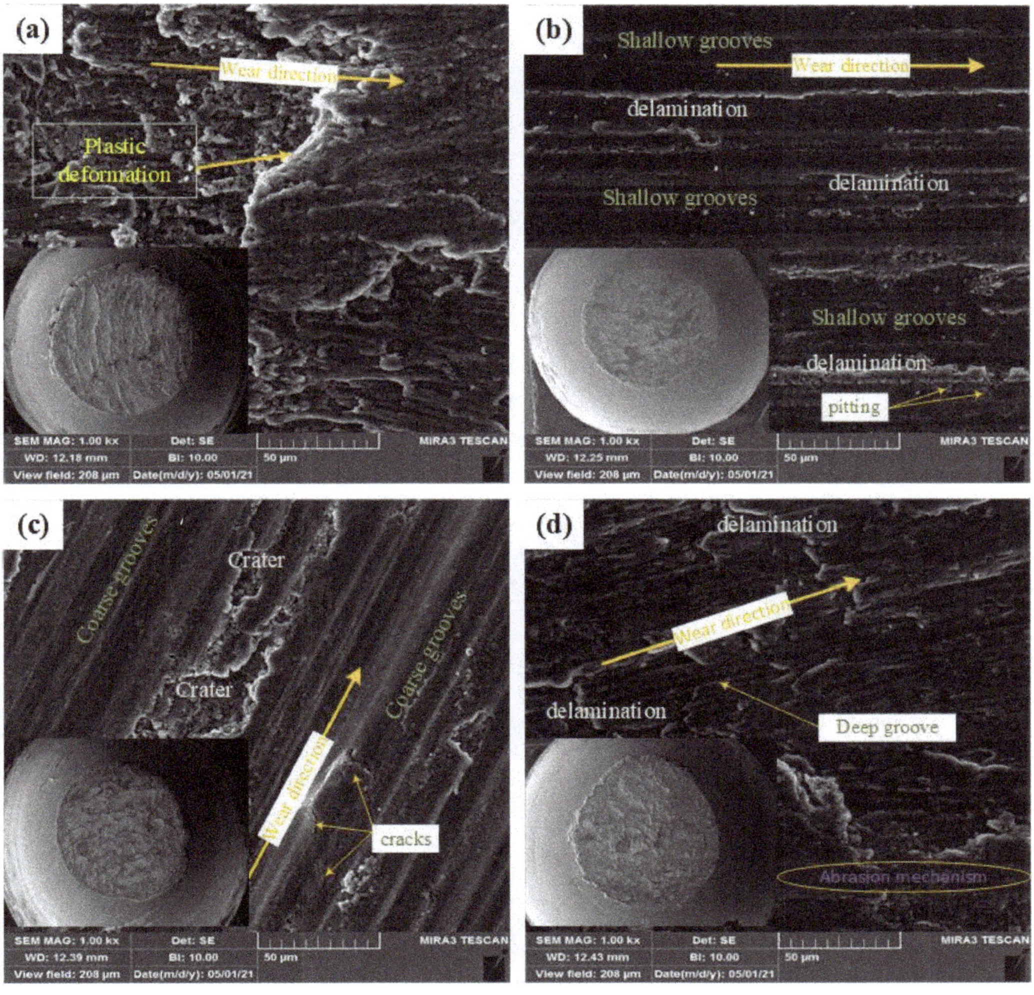

Figure 12. SEM images of the worn surfaces: (**a**) AMC0, (**b**) AMNC2, (**c**) AMNC4, and (**d**) AMNC6.

Figure 12b shows the worn surface of AMNC2 nanocomposite. The amount of flash generated in this sample is significantly reduced, probably owing to the hardness of the nanocomposite as well as the refined microstructure of the AMNC2 composite. Regular and relatively shallow grooves along the wear path, being the feature of scratch mechanism, are evident in this image. These grooves are created by the asperity of the opposite surface penetrating the surface of the pin and are mainly associated with the cutting and separation of ribbon-like chips from the surface. This mechanism often occurs in metal-to-metal contact when the pin surface is sufficiently hardened; however, if the subsurface layer is not strong enough, delamination is caused wherein microcracks are created which then connected each other parallel to the surface [5]. The delamination mechanism can also be seen in this sample, to some extent. To a small extent, pitting is also visible, which is a characteristic of the delamination mechanism. Therefore, the dominant mechanism in this composite is mainly abrasion and then, to some extent, delamination.

By adding more nanoparticles to the melt (i.e., twice the AMNC2), the wear mechanism of AMNC4 has not been considerably changed, as seen in Figure 12c. A combination of abrasion and delamination mechanisms is still detectable. The addition of further nanoparticles to the melt in AMNC4 led to an increased viscosity and hence a decreased ultrasonic treatment performance. With the increase in the porosity amount, the bonding strength between the reinforcing agent and matrix is reduced; therefore, the cracks can be initiated and grown as seen in some parts of the abraded surface. Probably, these cracks have spread and, in the parts with weaker bonding, relatively large pieces have separated and created pits on the surface. Studying the flash, it can be seen that there is no noticeable change between the reinforced composite samples. In other words, having the nanoparticles in the metal matrix, the amount of plastic deformation has been reduced and the developed flash has, in all likelihood, regularly been separated from the surface during the wear test.

Considering the issue of insufficient bonding of nanoparticles in higher weight fractions and the possibility of agglomerations, abrasion mechanism is seen on a small part of the AMNC6 surface and looks like islands in the middle of the entire surface (see Figure 12d). In addition, delamination is also recognizable with higher thicknesses. Deep grooves are probably created by the hard silica particles separated from the abraded surface to facilitate three-body wear. The aforesaid hard particles could be either those particles weakly bonded to the base matrix or the agglomerated phase that has been fragmented by wear forces.

An increase in the hardness of the nanocomposite may reduce the materials transfer from the pin surface to the disc surface. In other words, nanoparticles and the hardened structure of the pin surface tend to partially diffuse into the disc, facilitating the iron transfer to the pin. In long distances and increasing temperature, oxygen could be absorbed to form iron oxide as a tribolayer [53,72]. In Figure 13a–d, the wear debris of monolithic and composite samples reinforced with different weight fractions, along with the EDS results, are presented (Figure 13e–h). Seen in Figure 13, the wear debris of the unreinforced samples is greater, irregular in shape, thicker, and their deformed shapes are more pronounced. This type of chipping mostly occurs during the plastic deformation of the material, being a feature of adhesive mechanism. The size of the debris in the AMNC2 sample is relatively smaller, and their thickness has also been reduced. Some plate-like chips are also observed in AMNC2. This type of chipping is usually seen when material is moved at the edges of the wear grooves. In the samples AMNC4 and AMNC6, a combination of the all kinds of chips with irregular shape, sheet-like, and flaky morphology is observed. Figure 13 shows the EDS result of monolithic and A356/SiO_2 nanocomposite samples; Fe and O elements are detected.

These elements, combined with the temperature rise during the wear test, provide the conditions for the formation of oxidized wear. The strength and weakness of this mechanism are demonstrated by the different percentages of iron observed in the EDS analysis. For example, as shown in Figure 13e, the percentage of iron transferred to the surface of the pin is 3.55%, while this amount has reached about 28% in the AMNC2 sample. In samples with a higher weight fraction of nanosilica, the amount of transferred iron has been reduced by reducing the hardness. The amount of iron transferred from the disc surface to the pin surface is probably related to the hardness of the composite samples. In other words, by increasing the hardness of the samples, the penetration of the asperity of the pin surface to the disk increases and more iron is transferred to the pin. As the temperature rises, the conditions for the formation of an oxide layer are provided, preventing direct contact between metal and metal, and the amount of wear decreases [73,74].

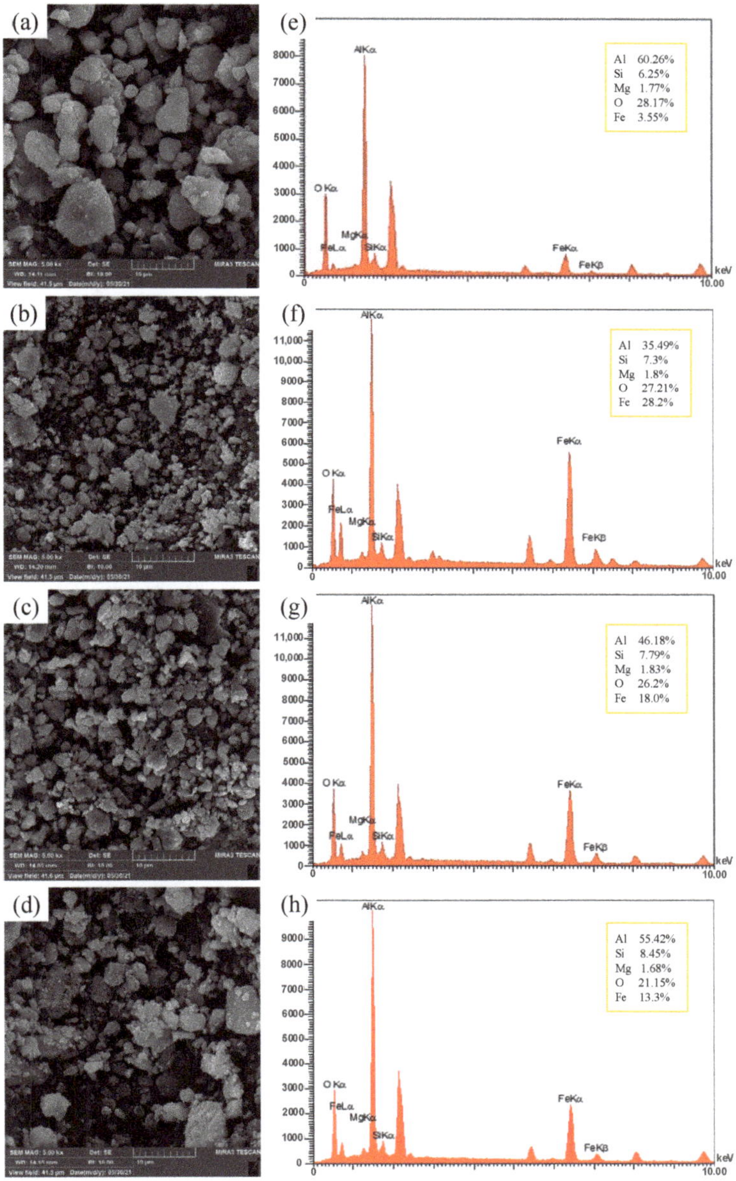

Figure 13. SEM image of debris: (**a**) unreinforced alloy, (**b**) AMNC2, (**c**) AMNC4, (**d**) AMNC6. The EDS of wear debris for the (**e**) unreinforced alloy, (**f**) AMNC2, (**g**) AMNC4, and (**h**) AMNC6.

4. Conclusions

The main conclusions of the present study are described as follows:

1. A proper ultrasonic melt treatment caused a better dispersion of the nanosilica reinforcements in the A356 aluminum alloy and refined the microstructure of the fabricated composite materials. Adding nanoparticles with 0.125 wt.% resulted in better wear resistance due to the synergistic improvement in both the reinforcing role of the nanosilica particles and ultrasonic efficiency. Adding more than this amount (0.125 wt.%), however, resulted in a relatively coarse microstructure due to the increase in viscosity and thus a decrease in the efficiency of ultrasonic waves;
2. Based on the hardness profile in the longitudinal direction of the cast ingot, the highest hardness is measured at the bottom and then at the top of the cast ingot. The hardness of all samples reinforced with nanosilica particles was higher than the base alloy. Among these, the largest increase is obtained from the A356/0.125% SiO_2 sample, exhibiting a 52% hardness increase;
3. The pin-on-disc experiment revealed that all the composite samples have a lower weight loss rate compared to the monolithic material. Investigations showed that the nanocomposites have better wear resistance at higher forces; for example, the A356/0.125% SiO_2 sample under the force of 100 N has recorded an improvement in wear resistance of about 68%;
4. As much as 50% improvement of the COF in the sample with the optimal wt.% (A356/0.125% SiO_2), and a reduction in the COF of all composite samples compared to the base alloy, is seen in this research. The COF in the composite samples did not change much with the increase in force, but the COF of the matrix alloy decreased with the addition of force;
5. Examining the abraded surfaces revealed that the dominant wear mechanisms in the samples reinforced with nanosilica are mainly abrasion and delamination. Further, the amount of abrasion in the A356/0.125% SiO_2 composite sample is higher and the destruction in this sample has reached its lowest level. EDS investigation of the wear debris showed that the iron transfer from the disc surface to the pin surface has taken place with the addition of nanosilica, and the iron transfer increased with higher hardness.

Author Contributions: All authors contributed to the study conception and design. Material preparation, data collection, and analysis were performed by all authors. The first draft of the manuscript was written by A.G. and all authors commented on previous versions of the manuscript. All authors have read and agreed to the published version of the manuscript.

Funding: This research received no external funding.

Data Availability Statement: The datasets generated and/or analyzed during the current study are available in this manuscript and there are no other data that are not in this manuscript.

Conflicts of Interest: The authors declare no conflict of interest.

References

1. Hassan, S.F.; Gupta, M. Development of high performance magnesium nano-composites using nano-Al_2O_3 as reinforcement. *Mater. Sci. Eng. A* **2005**, *392*, 163–168. [CrossRef]
2. Goh, C.S.; Wei, J.; Lee, L.C.; Gupta, M. Properties and deformation behaviour of Mg–Y_2O_3 nanocomposites. *Acta Mater.* **2007**, *55*, 5115–5121. [CrossRef]
3. Banerjee, S.; Poria, S.; Sutradhar, G.; Sahoo, P. Nano-indentation and corrosion characteristics of ultrasonic vibration assisted stir-cast AZ31–WC–graphite nano-composites. *Int. J. Met.* **2021**, *15*, 1058–1072. [CrossRef]
4. Alipour, M.; Keshavamurthy, R.; Koppad, P.G.; Shakiba, A.; Reddy, N.C. Investigation of microstructure and mechanical properties of cast Al–10Zn–3.5 Mg–2.5 Cu nanocomposite reinforced with graphene nano sheets produced by ultrasonic assisted stir casting. *Int. J. Met.* **2022**, *17*, 935–946.
5. Ghahremani, A.; Abdullah, A.; Arezoodar, A.F. Wear behavior of light-weight metal matrix nanocomposites. *Ceram. Int.* **2022**, *48*, 35947–35965. [CrossRef]

6. Yuvaraj, N.; Aravindan, S. Fabrication of Al5083/B_4C surface composite by friction stir processing and its tribological characterization. *J. Mater. Res.* **2015**, *4*, 398–410. [CrossRef]
7. Murthy, N.V.; Reddy, A.P.; Selvaraj, N.; Rao, C.S.P. Preparation of SiC based Aluminium metal matrix nano composites by high intensity ultrasonic cavitation process and evaluation of mechanical and tribological properties. In *IOP Conference Series: Materials Science and Engineering*; IOP Publishing: Bristol, UK, 2016; Volume 149. [CrossRef]
8. Venkatesh, R.; Rao, V.S. Thermal, corrosion and wear analysis of copper based metal matrix composites reinforced with alumina and graphite. *Def. Technol.* **2018**, *14*, 346–355. [CrossRef]
9. Mavhungu, S.; Akinlabi, E.; Onitiri, M.; Varachia, F. Aluminum matrix composites for industrial use: Advances and trends. *Procedia Manuf.* **2017**, *7*, 178–182. [CrossRef]
10. Xu, J.; Chen, L.; Choi, H.; Li, X. Theoretical study and pathways for nanoparticle capture during solidification of metal melt. *J. Phys. Condens. Matter* **2012**, *24*, 255304. [CrossRef]
11. Pai, B.C.; Ramani, G.; Pillai, R.M.; Satyanarayana, K.G. Role of magnesium in cast aluminium alloy matrix composites. *J. Mater. Sci.* **1995**, *30*, 1903–1911. [CrossRef]
12. Kobashi, M.; Choh, T. The wettability and the reaction for SiC particle/Al alloy system. *J. Mater. Sci.* **1993**, *28*, 684–690. [CrossRef]
13. Glass, S.J.; RolfWaesche, K.U.; Naito, M. Effects of titanium on wettability and interfaces in aluminum/ceramic systems. In *Characterization & Control of Interfaces for High Quality Advanced Materials*; The American Ceramic Society: Columbus, OH, USA, 2005; p. 83.
14. Nosko, M.; Nagy, Š.; Weber, L.; Mat'ko, I.; Mihalkovič, M.; Iždinský, K.; Orovčík, L'. Effect of Ca addition on interface formation in Al (Ca)/Al_2O_3 composites prepared by gas pressure assisted infiltration. *Mater. Des.* **2016**, *108*, 618–623. [CrossRef]
15. An, Q.; Cong, X.-S.; Shen, P.; Jiang, Q.-C. Roles of alloying elements in wetting of SiC by Al. *J. Alloys Compd.* **2019**, *784*, 1212–1220. [CrossRef]
16. Krishnan, P.K.; Arunachalam, R.; Husain, A.; Al-Maharbi, M. Studies on the influence of stirrer blade design on the microstructure and mechanical properties of a novel aluminum metal matrix composite. *J. Manuf. Sci. Eng.* **2021**, *143*, 021008. [CrossRef]
17. Karbalaei Akbari, M.; Baharvandi, H.R.; Mirzaee, O. Nano-sized aluminum oxide reinforced commercial casting A356 alloy matrix: Evaluation of hardness, wear resistance and compressive strength focusing on particle distribution in aluminum matrix. *Compos. Part B Eng.* **2013**, *52*, 262–268. [CrossRef]
18. Malaki, M.; Fadaei Tehrani, A.; Niroumand, B.; Gupta, M. Wettability in metal matrix composites. *Metals* **2021**, *11*, 1034. [CrossRef]
19. Malaki, M.; Tehrani, A.F.; Niroumand, B.; Abdullah, A. Ultrasonically Stir Cast SiO_2/A356 Metal Matrix Nanocomposites. *Metals* **2021**, *11*, 2004. [CrossRef]
20. Jebeen Moses, J.; Joseph Sekhar, S. Investigation on the tensile strength and microhardness of AA6061/TiC composites by stir casting. *Trans. Indian Inst. Met.* **2017**, *70*, 1035–1046. [CrossRef]
21. Bui, R.T.; Ouellet, R.; Kocaefe, D. A two-phase flow model of the stirring of Al-SiC composite melt. *Metall. Mater. Trans. B* **1994**, *25*, 607–618. [CrossRef]
22. Yamamoto, T.; Suzuki, A.; Komarov, S.V.; Ishiwata, Y. Investigation of impeller design and flow structures in mechanical stirring of molten aluminum. *J. Mater. Process. Technol.* **2018**, *261*, 164–172. [CrossRef]
23. Torotwa, I.; Ji, C. A study of the mixing performance of different impeller designs in stirred vessels using computational fluid dynamics. *Designs* **2018**, *2*, 10. [CrossRef]
24. Wu, J.; Pullum, L. Performance analysis of axial-flow mixing impellers. *AIChE J.* **2000**, *46*, 489–498. [CrossRef]
25. Mehta, V.R.; Sutaria, M.P. Investigation on the effect of stirring process parameters on the dispersion of SiC particles inside melting crucible. *Met. Mater. Int.* **2021**, *27*, 2989–3002. [CrossRef]
26. Naher, S.; Brabazon, D.; Looney, L. Computational and experimental analysis of particulate distribution during Al–SiC MMC fabrication. *Compos. Part A Appl. Sci. Manuf.* **2007**, *38*, 719–729. [CrossRef]
27. Abdullah, A.; Malaki, M.; Baghizadeh, E. On the impact of ultrasonic cavitation bubbles. *Proc. Inst. Mech. Eng. Part C J. Mech. Eng. Sci.* **2012**, *226*, 681–694. [CrossRef]
28. Abdullah, A.; Pak, A.; Abdullah, M.M.; Shahidi, A.; Malaki, M. Study of the behavior of ultrasonic piezo-ceramic actuators by simulations. *Electron. Mater. Lett.* **2014**, *10*, 37–42. [CrossRef]
29. Yang, Y.; Li, X. Ultrasonic cavitation-based nanomanufacturing of bulk aluminum matrix nanocomposites. *J. Manuf. Sci. Eng.* **2007**, *129*, 497–501. [CrossRef]
30. Cao, G.; Konishi, H.; Li, X. Recent developments on ultrasonic cavitation based solidification processing of bulk magnesium nanocomposites. *Int. J. Met.* **2008**, *2*, 57–65. [CrossRef]
31. Puga, H.; Barbosa, J.; Teixeira, J.C.; Prokic, M. A new approach to ultrasonic degassing to improve the mechanical properties of aluminum alloys. *J. Mater. Eng. Perform.* **2014**, *23*, 3736–3744. [CrossRef]
32. Tekmen, C.; Saday, F.; Cocen, U.; Ljungberg, L.Y. An investigation of the effect of SiC reinforcement coating on the wettability of Al/SiC system. *J. Compos. Mater.* **2008**, *42*, 1671–1679. [CrossRef]
33. Feng, A.; McCoy, B.J.; Munir, Z.A.; Cagliostro, D. Wettability of transition metal oxide surfaces. *Mater. Sci. Eng. A* **1998**, *242*, 50–56. [CrossRef]
34. Eskin, G.I. *Ultrasonic Treatment of Light Alloy Melts*; CRC Press: Boca Raton, FL, USA, 1998.
35. Malaki, M.; Xu, W.; Kasar, A.K.; Menezes, P.L.; Dieringa, H.; Varma, R.S.; Gupta, M. Advanced metal matrix nanocomposites. *Metals* **2019**, *9*, 330. [CrossRef]

36. Chen, L.-Y.; Xu, J.-Q.; Choi, H.; Pozuelo, M.; Ma, X.; Bhowmick, S.; Yang, J.-M.; Mathaudhu, S.; Li, X.-C. Processing and properties of magnesium containing a dense uniform dispersion of nanoparticles. *Nature* **2015**, *528*, 539–543. [CrossRef] [PubMed]
37. Eskin, G.I.; Eskin, D.G. Production of natural and synthesized aluminum-based composite materials with the aid of ultrasonic (cavitation) treatment of the melt. *Ultrason. Sonochemistry* **2003**, *10*, 297–301. [CrossRef]
38. Tzanakis, I.; Xu, W.W.; Eskin, D.G.; Lee, P.D.; Kotsovinos, N. In situ observation and analysis of ultrasonic capillary effect in molten aluminium. *Ultrason. Sonochemistry* **2015**, *27*, 72–80. [CrossRef]
39. Liu, X.; Jia, S.; Nastac, L. Ultrasonic cavitation-assisted molten metal processing of cast A356-nanocomposites. *Int. J. Met.* **2014**, *8*, 51–58. [CrossRef]
40. Gallego-Juárez, J.A.; Graff, K.F. *Power Ultrasonics: Applications of High-Intensity Ultrasound*; Elsevier: Amsterdam, The Netherlands, 2014.
41. Donthamsetty, S.; Suresh Babu, P. Experiments on the wear characteristics of A356 MMNCs fabricated using ultrasonic cavitation. *Int. J. Automot. Mech. Eng.* **2017**, *14*, 4589–4602. [CrossRef]
42. Hashim, J.; Looney, L.; Hashmi, M.S.J. The wettability of SiC particles by molten aluminium alloy. *J. Mater. Process. Technol.* **2001**, *119*, 324–328. [CrossRef]
43. Jayashree, P.K.; Shankar, M.C.G.; Kini, A.; Sharma, S.S.; Shetty, R. Review on effect of silicon carbide (SiC) on stir cast aluminium metal matrix composites. *Int. J. Curr. Eng. Technol.* **2013**, *3*, 1061–1071.
44. Abedini, R.; Abdullah, A.; Alizadeh, Y. Ultrasonic assisted hot metal powder compaction. *Ultrason. Sonochemistry* **2017**, *38*, 704–710. [CrossRef]
45. Sarvi Hampa, P.; Razfar, M.R.; Malaki, M.; Maleki, A. The role of dry aero-acoustical lubrication and material softening in ultrasonically assisted milling of difficult-to-cut AISI 304 steels. *Trans. Indian Inst. Met.* **2015**, *68*, 43–49. [CrossRef]
46. Malaki, M.; Ding, H. A review of ultrasonic peening treatment. *Mater. Des.* **2015**, *87*, 1072–1086. [CrossRef]
47. Abdullah, A.; Malaki, M.; Eskandari, A. Strength enhancement of the welded structures by ultrasonic peening. *Mater. Des.* **2012**, *38*, 7–18. [CrossRef]
48. Shafiei-Zarghani, A.; Kashani-Bozorg, S.F.; Hanzaki, A.Z. Wear assessment of Al/Al_2O_3 nano-composite surface layer produced using friction stir processing. *Wear* **2011**, *270*, 403–412. [CrossRef]
49. Hosseini, N.; Karimzadeh, F.; Abbasi, M.H.; Enayati, M.H. A comparative study on the wear properties of coarse-grained Al6061 alloy and nanostructured Al6061–Al_2O_3 composites. *Tribol. Int.* **2012**, *54*, 58–67. [CrossRef]
50. Alam, M.T.; Arif, S.; Ansari, A.H. Wear behaviour and morphology of stir cast aluminium/SiC nanocomposites. *Mater. Res. Express* **2018**, *5*, 045008. [CrossRef]
51. Sharma, A.; Fujii, H.; Paul, J. Influence of reinforcement incorporation approach on mechanical and tribological properties of AA6061- CNT nanocomposite fabricated via FSP. *J. Manuf. Process.* **2020**, *59*, 604–620. [CrossRef]
52. Suresh, R.; Siddeshkumar, N. Investigation on dry sliding wear behavior of AA5083/nano-Al_2O_3 metal matrix composites. *Rev. Metal.* **2022**, *58*, e213. [CrossRef]
53. Karbalaei Akbari, M.; Rajabi, S.; Shirvanimoghaddam, K.; Baharvandi, H. Wear and friction behavior of nanosized TiB_2 and TiO_2 particle-reinforced casting A356 aluminum nanocomposites: A comparative study focusing on particle capture in matrix. *J. Compos. Mater.* **2015**, *49*, 3665–3681. [CrossRef]
54. George, S.M.; Priya, R.; Vijayakumar, G.N.S.; Pradeep, J.A. Study on mechanical characteristics of the nano-TiC reinforced Al6061 metal matrix composites. *Mater. Today Proc.* **2022**, *62*, 2224–2229. [CrossRef]
55. Miyajima, T.; Iwai, Y. Effects of reinforcements on sliding wear behavior of aluminum matrix composites. *Wear* **2003**, *255*, 606–616. [CrossRef]
56. Hosseini, N.; Karimzadeh, F.; Abbasi, M.; Enayati, M.H. Tribological properties of Al6061–Al_2O_3 nanocomposite prepared by milling and hot pressing. *Mater. Des.* **2010**, *31*, 4777–4785. [CrossRef]
57. Abdollahi, A.; Alizadeh, A.; Baharvandi, H.R. Dry sliding tribological behavior and mechanical properties of Al2024–5 wt.% B_4C nanocomposite produced by mechanical milling and hot extrusion. *Mater. Des.* **2014**, *55*, 471–481. [CrossRef]
58. Hutchings, I.; Shipway, P. *Tribology: Friction and Wear of Engineering Materials*; Butterworth-Heinemann: Oxford, UK, 2017.
59. Saba, F.; Zhang, F.; Liu, S.; Liu, T. Reinforcement size dependence of mechanical properties and strengthening mechanisms in diamond reinforced titanium metal matrix composites. *Compos. Part B Eng.* **2019**, *167*, 7–19. [CrossRef]
60. Ramesh, C.S.; Ahamed, A. Friction and wear behaviour of cast Al 6063 based in situ metal matrix composites. *Wear* **2011**, *271*, 1928–1939. [CrossRef]
61. Fathy, A.; Shehata, F.; Abdelhameed, M.; Elmahdy, M. Compressive and wear resistance of nanometric alumina reinforced copper matrix composites. *Mater. Des. (1980–2015)* **2012**, *36*, 100–107. [CrossRef]
62. Sadoun, A.M.; Fathy, A.; Abu-Oqail, A.; Elmetwaly, H.T.; Wagih, A. Structural, mechanical and tribological properties of Cu–ZrO_2/GNPs hybrid nanocomposites. *Ceram. Int.* **2020**, *46*, 7586–7594. [CrossRef]
63. Kerr, I.; Priest, M.; Okamoto, Y.; Fujita, M. Friction and wear performance of newly developed automotive bearing materials under boundary and mixed lubrication regimes. *Proc. Inst. Mech. Eng. Part J J. Eng. Tribol.* **2007**, *221*, 321–331. [CrossRef]
64. Guo, Q.B.; Rong, M.Z.; Jia, G.L.; Lau, K.T.; Zhang, M.Q. Sliding wear performance of nano-SiO_2/short carbon fiber/epoxy hybrid composites. *Wear* **2009**, *266*, 658–665. [CrossRef]
65. Zhang, G.; Chang, L.; Schlarb, A.K. The roles of nano-SiO_2 particles on the tribological behavior of short carbon fiber reinforced PEEK. *Compos. Sci. Technol.* **2009**, *69*, 1029–1035. [CrossRef]

66. Moazami-Goudarzi, M.; Akhlaghi, F. Wear behavior of Al 5252 alloy reinforced with micrometric and nanometric SiC particles. *Tribol. Int.* **2016**, *102*, 28–37. [CrossRef]
67. Rao, R.N.; Das, S.; Mondal, D.P.; Dixit, G. RETRACTED: Dry sliding wear behaviour of cast high strength aluminium alloy (Al–Zn–Mg) and hard particle composites. *Wear* **2009**, *267*, 1688–1695. [CrossRef]
68. Lakshmipathy, J.; Kulendran, B. Reciprocating wear behavior of 7075Al/SiC in comparison with 6061Al/Al_2O_3 composites. *Int. J. Refract. Met. Hard Mater.* **2014**, *46*, 137–144. [CrossRef]
69. Krishnan, P.; Lakshmanan, P.; Palani, S.; Arumugam, A.; Kulothungan, S. Analyzing the hardness and wear properties of SiC and hBN reinforced aluminum hybrid nanocomposites. *Mater. Today Proc.* **2022**, *62*, 566–571. [CrossRef]
70. Alipour, M.; Eslami-Farsani, R. Synthesis and characterization of graphene nanoplatelets reinforced AA7068 matrix nanocomposites produced by liquid metallurgy route. *Mater. Sci. Eng. A* **2017**, *706*, 71–82. [CrossRef]
71. Almomani, M.A.; Hayajneh, M.T.; Alelaumi, S.M. Applying Taguchi method to study the wear behaviour of ZA-27 alloy-based composites reinforced with SiC nanoparticles. *Int. J. Cast Met. Res.* **2019**, *32*, 229–241. [CrossRef]
72. Dinaharan, I.; Murugan, N.; Thangarasu, A. Development of empirical relationships for prediction of mechanical and wear properties of AA6082 aluminum matrix composites produced using friction stir processing. *Eng. Sci. Technol. Int. J.* **2016**, *19*, 1132–1144. [CrossRef]
73. Sharifi, E.M.; Karimzadeh, F. Wear behavior of aluminum matrix hybrid nanocomposites fabricated by powder metallurgy. *Wear* **2011**, *271*, 1072–1079. [CrossRef]
74. Yuvaraj, N.; Aravindan, S.; Vipin. Wear Characteristics of Al5083 Surface Hybrid Nano-composites by Friction Stir Processing. *Trans. Indian Inst. Met.* **2016**, *70*, 1111–1129. [CrossRef]

crystals

MDPI

Article

Influence of TiC Particles on Mechanical and Tribological Characteristics of Advanced Aluminium Matrix Composites Fabricated through Ultrasonic-Assisted Stir Casting

Chitti Babu Golla [1], Mahammod Babar Pasha [1], Rajamalla Narasimha Rao [1,*], Syed Ismail [1] and Manoj Gupta [2,*]

[1] Department of Mechanical Engineering, National Institute of Technology Warangal, Warangal 506004, India
[2] Department of Mechanical Engineering, National University of Singapore, 9 Engineering Drive 1, Singapore 117576, Singapore
* Correspondence: rnrao@nitw.ac.in (R.N.R.); mpegm@nus.edu.sg (M.G.)

Abstract: The present investigation highlights the development of high-performance materials in the construction materials industry, with a special focus on the production of aluminium matrix composites (AMCs) containing titanium carbide (TiC) particles. The stir casting method with ultrasonic assistance was employed to enhance the mechanical and tribological properties. ASTM standards were employed to evaluate the influence of TiC particles on density, hardness (VHN), ultimate tensile strength (UTS), and wear resistance at various TiC weight fraction percentages (0.0 wt.%, 2.0 wt.%, 4.0 wt.%, 6.0 wt.%, and 8.0 wt.%). Field emission scanning electron microscopy (FESEM) and X-ray diffraction (XRD) analysis were performed to analyse the microstructural changes and elemental phases present in the synthesised composite. Results revealed that the incorporation of 8 wt.% TiC reinforcement in the metal matrix composites demonstrated significant improvements compared to the base alloy. In particular, a substantial enhancement in hardness by 32%, a notable increase of 68% in UTS, and a significant 80% rise in yield strength were observed when contrasted with the pure aluminium alloy. The tensile fracture analysis of the specimens revealed the presence of dimples, voids, and cracks, suggesting a brittle nature. To assess the wear characteristics of the composites, dry sliding wear experiments were performed using a pin-on-disc wear tester. Incorporation of TiC particles resulted in a lower coefficient of friction than the base alloy, with the lowest friction coefficient being recorded at 0.266 for 6 wt.% TiC, according to the data. FESEM and energy-dispersive X-ray spectroscopy (EDXS) were used to examine the surfaces of the worn pin. Overall, the inclusion of TiC reinforcement particles in the matrix alloy greatly enhanced the wear resistance and friction coefficient of the Al-6TiC composites. Ploughing and adhesion under lower loads and delamination under higher loads were the wear mechanisms observed in the wear test.

Keywords: aluminium; TiC; ultrasonic-assisted stir casting; wear resistance; wear mechanism

1. Introduction

In recent decades, metal matrix composites (MMCs) have emerged as new engineering materials with superior properties compared to monolithic materials. The favourable properties that are typically exhibited by metal matrix composites include light weight, high elastic modulus, high strength, high thermal stability, resistance to elevated temperatures, improved wear resistance, a low thermal expansion coefficient, and extended fatigue life [1,2]. Due to their exceptional blend of properties, advanced matrix composites (AMCs) have risen to prominence as advanced materials, playing a crucial role in diverse industries such as aerospace, automotive [3] (i.e., pistons and connecting rod [4]), defence, and various engineering sectors [5]. Aluminium [6–8], and magnesium are the usual matrices used for lightweight MMCs [9]. The most commonly used aluminium alloy (AA) matrix ranges from the AA-1xxx series to the AA-8xxx series. Among these, the AA-6xxx series and AA-7xxx series alloys are broadly researched because of their excellent corrosion resistance, wear resistance, and high strength. Recently, AA8011 has become the preferred material for structural applications in modern engineering, including construction and transportation. This is due to its composition of Fe and Si alloying elements, which provide the necessary characteristics of good hardness, high tensile strength, and suitable wear resistance, which are crucial for such applications [10]. Despite its favourable properties, its use in automobile powertrain applications is limited by its poor wear resistance, which necessitates further improvement through the development of advanced composites [11]. To improve the low wear resistance of AA8011, researchers have studied the integration of hard secondary reinforcement phases into the aluminium matrix. Reinforcements such as carbides, borides, and oxides have been utilised for both ex situ and in situ composites [12]. The mechanical and surface properties of 8000 series aluminium alloys were enhanced through the incorporation of ceramic particles, such as Al_2O_3 [13], SiC [14,15], TiO_2 [16], Si_3N_4/ZrB_2 [17], B_4C [18] and TiB_2 [19]. AMCs are presently manufactured using different methods such as mechanical alloying, additive manufacturing [20,21], stir casting [22], compo casting [23], powder metallurgy [24], squeeze casting [25], and spray deposition [26].

Liquid state processes have proven to be the most economical and productive approach for fabricating metal matrix composites. Their straightforward utilisation, economical processing cost, ability to scale up to large-scale production, and high output render them a preferred option. Stir casting, a typical liquid-state processing method for generating AMCs, has been combined with ultra-sonication in recent studies to produce efficient cavitation and ensure an even dispersion of reinforcement particles throughout the matrix [27]. Sakthivelu et al. [28] developed AA8011/Al_2O_3 composites by changing the weight percentage (4, 8, 12 wt.%) of the Al_2O_3 and investigated the composites' sliding wear behaviour. According to the results, composites were significantly more wear-resistant than base alloys. Siddharthan et al. [29] reported that SiC particle-reinforced AA8011 composites displayed improved resistance to wear. Likewise, Fayomi et al. [30] produced an AA8011 alloy composite with changing weight fractions of nano-Si_3N_4 using a two-step stir casting method. The authors concluded that the gradual presence of Si_3N_4 particles reduced the friction coefficient while increasing the resistance to wear. Moreover, associated with the base material, the hybrid AA8011 composites with Si_3N_4 and ZrB_2 reinforcements displayed much higher wear resistance. Keerthivasan et al. [31] demonstrated that the inclusion of TiC particles significantly enhances the mechanical behaviour of the composites, resulting in improvements in hardness and ultimate tensile strength. Baskaran et al. [32] observed that the integration of TiC particles into the AA7075 matrix leads to a substantial enhancement in wear resistance. Similarly, Siddappa et al. [33] developed AMCs based on Al6061 and TiC particulates (3%, 6%, 9%, and 12%) using the stir casting process. The results revealed a significant improvement in the hardness and tensile strength of the composites, with an enhancement of approximately 20% observed as the TiC particulate content progressively increased.

A comprehensive investigation has been carried out on various aluminium alloy metal matrix composites, including AA6061 [34], AA7075 [35], and AA2024 [36,37], as evident from the literature. However, there exists a limited amount of research focusing on the microstructural analysis, hardness, and wear characteristics of TiC-reinforced AA8011 matrix composites. As a result, the purpose of this research is to investigate how the presence of TiC particles affects the microstructure, hardness, and wear properties of AA8011 matrix composites. By addressing this research gap, valuable insights can be gained regarding the potential applications and performance of TiC-reinforced AA8011 composites. The outcomes of this investigation are likely to offer a deeper understanding of the material behaviour and pave the way for optimising the properties of AMCs in various engineering applications.

2. Materials and Methods

2.1. Materials

The present investigation utilised aluminium alloy (AA8011) as the base material (Table 1). It was procured from Bharat Aerospace Metals Pvt Ltd., Mumbai, India. Properties of AA8011 include density (2.7 g/cm^3), ultimate tensile strength (124 MPa), melting point (660 °C), modulus of elasticity (70.3 GPa), and thermal conductivity (237 W/m·k) [10].

Table 1. Chemical elements present in AA8011 aluminium alloy matrix (in wt.%).

Element	Si	Fe	Ag	Mn	Cu	Mg	Ti	Zn	Al
AA8011	0.624	0.721	0.45	0.04	0.23	0.30	0.003	0.002	Bal

The TiC particles having an average particle size of 10 μm were procured from Parshwamani Metals, Mumbai, India. TiC particles have gained significant attention as reinforcement in AMCs due to their outstanding features, including a high melting temperature (3250 °C), high modulus of elasticity (450 GPa), high hardness (2470 Kg/mm^2), a density of 4.93 g/cm^3, a low coefficient of thermal expansion (7.61 μm/k), and excellent thermal conductivity (28.9 W/m·k). The TiC particle composition was revealed through a scanning electron micrograph (Figure 1a), showcasing particles with irregular shapes and various sizes. This variability in particle attributes arose due to the ex situ manufacturing process, which lacked precise control over size and shape. In (Figure 1b), the X-ray diffraction pattern of TiC is presented, exhibiting distinct peaks at (1 1 1), (2 0 0), (2 2 0), (1 1 3), and (2 2 2). These peaks confirm the nature of the recovered powder as TiC particles.

Figure 1. (**a**) Scanning electron micrograph of TiC particles. (**b**) X−ray diffraction analysis of TiC particles.

2.2. Fabrication of the AA8011-TiC Composites

In this study, an ultrasonically aided stir casting process was used to modify a traditional stir casting process for developing metal matrix composites. The AA8011 aluminium alloy matrix was melted in an electrical furnace at around 800 °C, which is higher than the melting temperature of pure aluminium. During the melting process, a mechanical stirrer was used to stir the molten metal. To improve the wettability of the AA8011 matrix, magnesium ribbons were used between the matrix and the TiC reinforcement. The molten surface was completely cleansed, and the molten metal's temperature was then raised to 750 °C. Preheated TiC microparticles were introduced to the liquid metal at various weight ratios, resulting in the formation of composite materials with different compositions. The stirring process was carefully maintained to ensure a homogeneous dispersion, with an average stirring speed of 350 rpm for approximately 15 min [38]. Afterward, the composite mixture was allowed to cool and reach a semi-solid state for 15 min. It was then reheated to 750 °C and subjected to an additional 5 min of stirring. The stirring speed, which was controlled between 300 and 350 rpm, was constantly monitored. Ultrasonic-assisted stirring was used at this step, with a high-power ultrasonic generator with an input frequency of 20 kHz and an input power of 2 kW. An ultrasonic probe was placed in the composite blend for around 15 min, producing high-intensity ultrasonic sound waves [39]. Many dendrites with many branches might grow in the casting during the solidification process, reducing the mechanical properties of the composite. In order to remove this type of dendritic structure, an ultrasonic vibration was used during casting. The vibrations assist in breaking the dendritic structure and evenly distribute the particles in the casting [40]. This also enhances wettability and uniform dispersion of the TiC particles within the matrix material. Figure 2a–c, depicts Composite fabrication setup and the use of ultrasonic-assisted stir casting (Swap Equip, Chennai, India) for preparing composites with varying weight percentages of TiC (0%, 2%, 4%, 6%, and 8%). The same process was repeated to create composites with varying wt.% of TiC particles. The as-cast composites were machined into wear-testing shapes.

Figure 2. Composite fabrication setup: (**a**) Stir casting setup, (**b**) Ultrasonic setup attached to the stir casting, (**c**) Fabricated composite samples.

To obtain a fine surface finish, the machined samples were mechanically ground using silicon carbide abrasive grit papers ranging in grit size from 800 to 2000 and polished with diamond paste. To investigate the microstructure of the polished samples, etching was conducted using Keller's reagent, and observation was carried out using SEM and EDS. The samples' hardness was measured using a Vickers microhardness test as per ASTM E384-17 standard. The test involved creating 20 indentations spaced 0.5 mm apart and applying a 100 gf load for 15 s [12]. To determine the microhardness value of the samples, an average was calculated based on the 20 measurements obtained.

As illustrated in Figure 3, the dry sliding wear testing was carried out utilising a pin-on-disc testing device made by MAGNUM Instruments Pvt. Ltd., Bangalore, India. The testing procedure followed the established protocols outlined in the ASTM G99-09 standard. The cylindrical pin-shaped samples measured 10 mm in diameter and 30 mm in length. They are intended to slide against a counter disc made of EN31 steel, a common material in the manufacture of mechanical components. The Rockwell hardness scale C specifies EN31 steel as having a hardness of 60 ± 3 HRC. As a result, the pin samples were designed to be subjected to a sliding test against a material of a specific hardness. Experiments were conducted in a room temperature environment with a constant sliding distance of 1000 m while the applied force and sliding speed were varied. The load levels included 10, 20, and 30 N [41], while the sliding speeds ranged from 1 to 5 m/s. The pin and counter disc surfaces were prepared for each test by polishing them with silicon carbide abrasive grit papers with different grades (600, 1000, and 1500). Following this, the surfaces were cleaned with acetone to ensure their cleanliness. Before each wear test, the average roughness value of both the pin and counter disc surfaces was maintained at approximately 0.5 microns. The weight loss method, described in Equations (1)–(3) [42,43], was used to determine the volumetric wear rate, and the coefficient of friction was estimated by averaging the coefficient across a 1000 m sliding distance. Each wear test was repeated three times to minimise experimental errors, and the average values were reported. The worn pin surfaces were observed using SEM equipped with EDS.

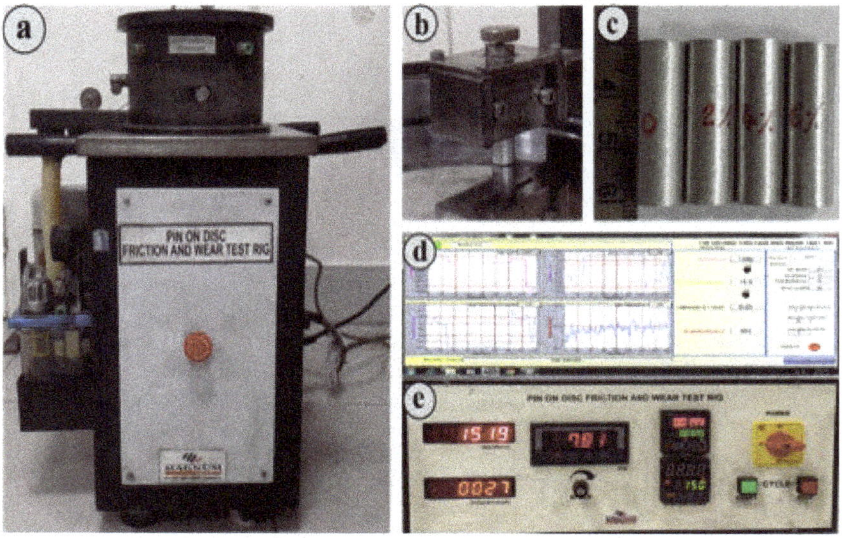

Figure 3. Dry sliding wear test: (**a**) Pin-on-disc setup, (**b**) a close-up of the pin holder, (**c**) pin samples, (**d**) real-time friction and wear graphs, and (**e**) machine controller.

$$\text{Mass loss of sample} = \frac{[\text{Initial Weight (g)} \; - \; \text{Final Weight (g)}]}{[\text{Density of the sample (g. mm}^{-3})]}\,\text{mm}^3 \tag{1}$$

$$\text{Volumetric wear rate} = \frac{[\text{Mass loss of the sample}]}{[\text{Sliding distance (m)}]}\left(\frac{\text{mm}^3}{\text{m}}\right) \tag{2}$$

$$\text{Coefficient of Friction } (\mu) = \frac{[\text{Frictional force}]}{[\text{Applied load}]} \tag{3}$$

3. Results

3.1. Microstructure Analysis of AA8011-TiC Composite

The optical micrographs for the AA8011-TiC composites are shown in Figure 4a–d, illustrating different weight percentages of TiC. (Figure 4a–b) shows the microstructure of the as-cast AA8011-2 wt.% TiC, and the 4 wt.% TiC composite, revealing a relatively uniform dispersion of particulates. Although the presence of TiC particles is less prominent compared to the higher TiC content samples, the introduction of reinforcement contributes to the refinement of the microstructure. Notably, the size of the coarse grain in AA8011 is reduced, indicating an improved microstructural refinement. Moving to (Figure 4c–d), the microphotographs present AA8011-TiC composites containing 6% and 8% TiC particles, respectively. These images demonstrate the consistent expansion of the ash-coloured region with increasing TiC concentration. No clustering or micro-pores were observed, indicating a homogeneous dispersion of TiC particles. Furthermore, increasing TiC content from 0% to 8% leads to a refinement of grain size from coarse to fine, consistent with previous studies [44]. The microstructure of the aluminium casting alloy with high titanium content is defined by the presence of dendrites during its solidification. In the spaces between these dendrites, one finds a composition of fine TiC particles and laths, originating from the eutectic mixture. As the molten mixture with high titanium content and TiC particles solidifies, the TiC particles are expelled towards the solid–liquid boundary and gather in the regions among the dendrites, forming the interdendritic areas. Consequently, this leads to a smaller grain size compared to castings without particles, and the TiC particles mix into the eutectic microstructure alongside Ti particles and laths [45].

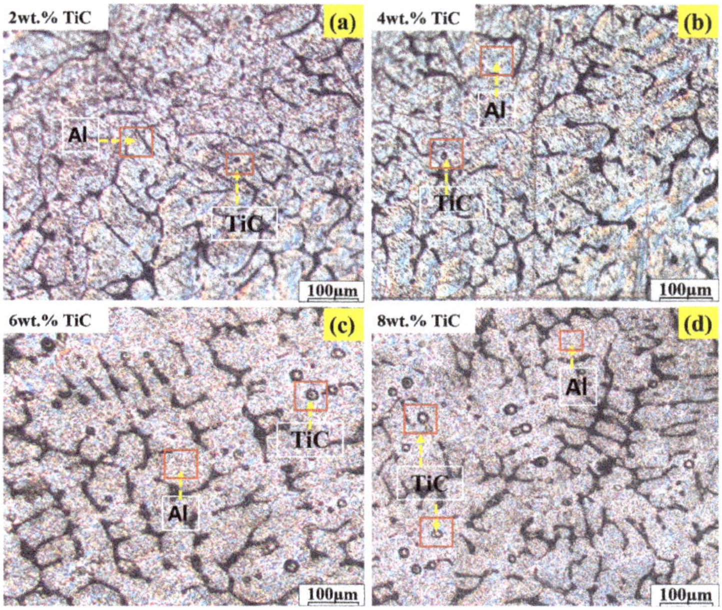

Figure 4. Optical microscope images of AA8011 alloy reinforced with varying amounts of TiC particles: (**a**) Al-2 wt.% TiC (**b**) Al-4 wt.% TiC, (**c**) Al-6 wt.% TiC, (**d**) Al-8 wt.% TiC.

Continuing the analysis to Figure 5, Figure 5a,b, presents SEM micrographs showing the as-cast AA8011 composites. The micrographs of the composites reveal a dendritic-type grain structure and the existence of TiC particles. The EDS spectrum of AA8011-2 wt.% TiC further confirms the existence of TiC particles within the composite material. Moving to Figure 5c,d, the microstructure of the TiC composite displays a dendritic morphology [46]. The inclusion of TiC particles contributes to the refinement of the grain structure, reducing the size of the large dendrites in the 4 wt.% TiC composite [47]. The EDS mapping of the matrix in Figure 5e,f, indicates the absence of TiC particles, whereas the 6 wt.% TiC composite samples reveal the presence of Ti and C elements. Furthermore, there are no signs of TiC particle aggregation in the composite samples, which can be attributed to the best selection of ultrasonic-assisted stir casting process parameters. Figure 5g,h, reveals the most significant improvement, where the dendritic microstructure is finely refined due to the higher number of individual TiC particles present in the 8 wt.% TiC sample. Figure 6a,b shows the spatial distribution and Elemental mapping of TiC in the 8 wt.% TiC composite sample. The results of this study highlight the ability of the processing methodology used in this study to ensure a reasonably uniform dispersion of TiC particles within the matrix.

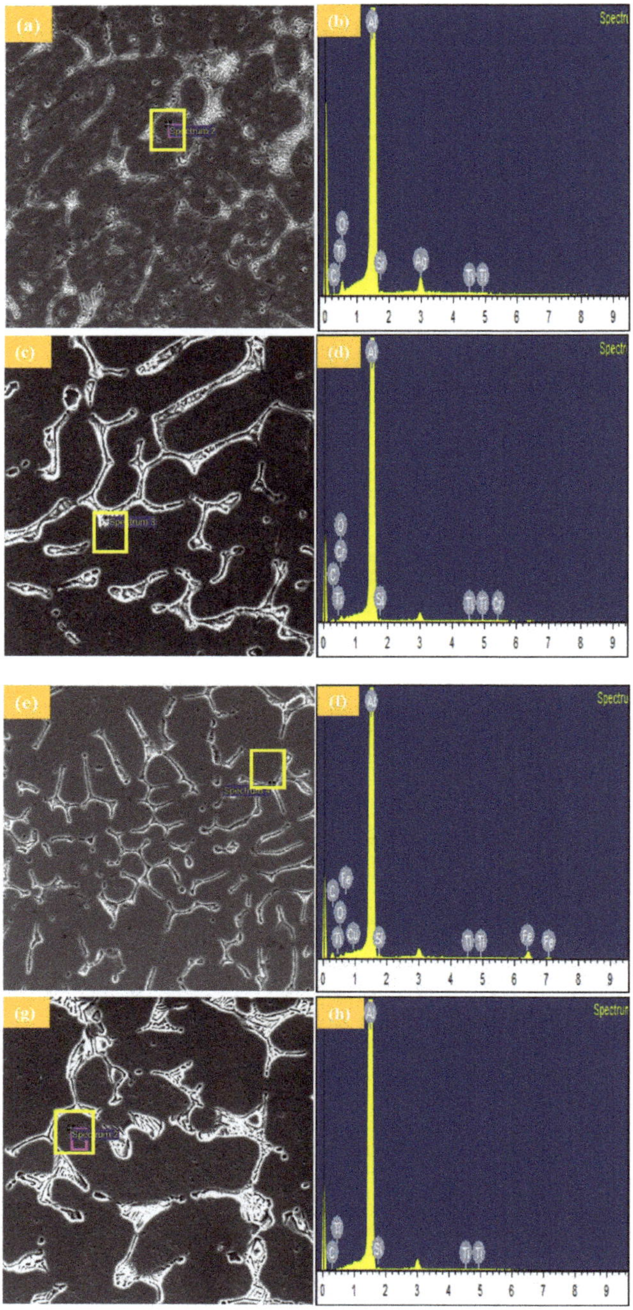

Figure 5. Scanning electron micrographs along with EDS mapping of the composites containing: (**a**,**b**) Al-2 wt.% TiC, (**c**,**d**) Al-4 wt.% TiC, (**e**,**f**) Al-6 wt.% TiC, (**g**,**h**) Al-8 wt.% TiC.

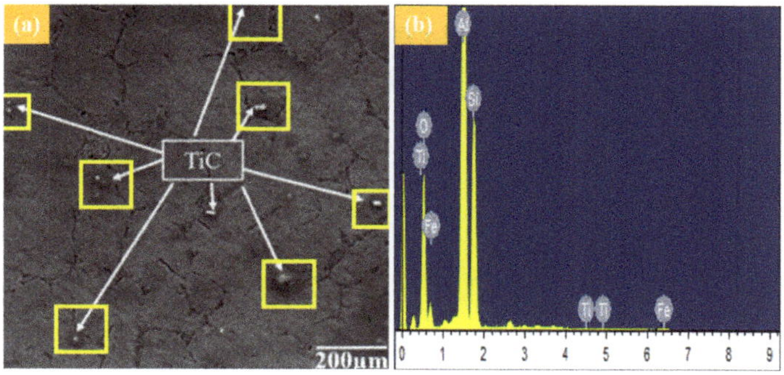

Figure 6. (**a**) SEM images depicting the spatial distribution of TiC particles in Al-8 wt.% TiC, (**b**) Elemental mapping showing the presence of major elements.

3.2. *Analysis of AA8011-TiC Composite Materials Using X-ray Diffraction*

The X-ray diffraction (XRD) pattern of the AA8011 aluminium alloy reinforced with TiC particulates is presented in Figure 7. The XRD pattern shows Al-TiC composites synthesised with different weight percentages (TiC 0–8 wt.%). The presence of aluminium is confirmed by the diffraction peaks observed at 38.55, 44.35, 65.30, and 78.35, corresponding to the Miller indices (111), (200), (220), and (311). The height of the diffraction peak in the XRD pattern for TiC is directly proportional to the content of ex situ-formed TiC particles in the composite material [48]. Therefore, increasing the number of TiC particles results in a corresponding increase in the peak height of the XRD pattern. In summary, in the XRD pattern of the Al-TiC composites the relationship between the peak height and the particulate content demonstrates the efficacy of the fabrication process in reinforcing the aluminium alloy with TiC particulates. Moreover, the absence of intermetallic compounds in the XRD pattern suggests that the resulting composite materials possess desirable properties suitable for various industrial applications [49].

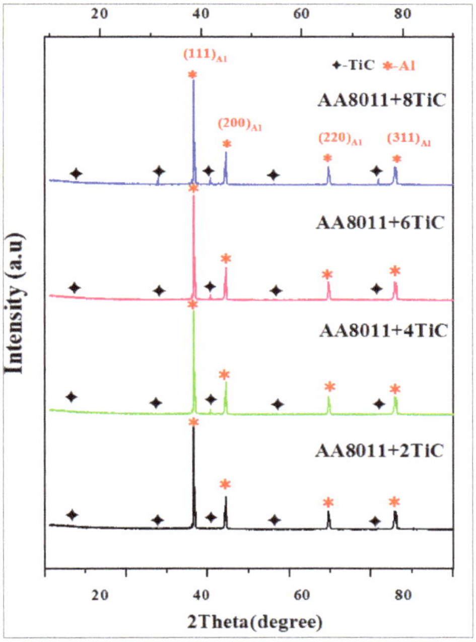

Figure 7. Results of X-ray diffraction on AA8011 as—cast and composites.

3.3. Micro—Hardness

Figure 8 shows the hardness measurement results. The base alloy AA8011 sample showed an average microhardness value of ~55.4 Vickers hardness (HV). With the incorporation of TiC particles, the hardness value was enhanced to 65.5 HV for the 2 wt.% TiC sample, ~72.46 HV for the 4 wt.% TiC sample, ~76.82 HV for 6 wt.% TiC sample, and a maximum of 80.2 HV for the 8 wt.% TiC specimen. The well-dispersed, harder TiC particles and their exceptional resistance to localised plastic deformation during indentation are the primary factors contributing to the significant improvement in microhardness. The microhardness value of the Al-TiC samples was much greater than the base alloy [50].

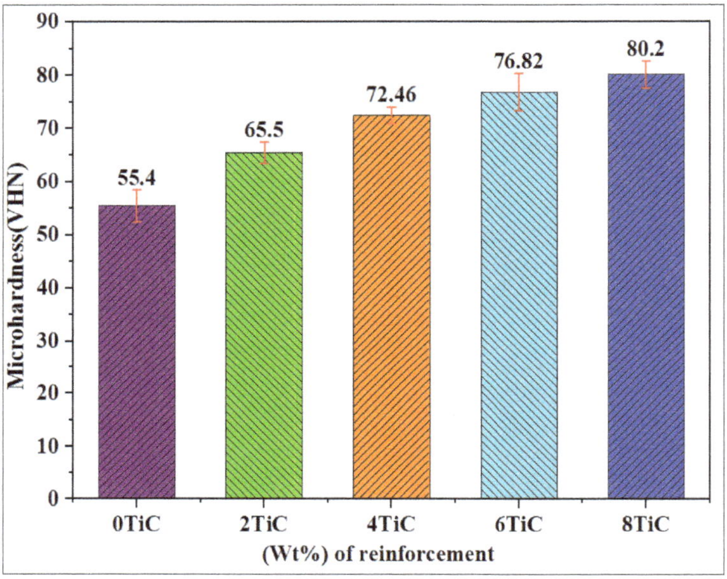

Figure 8. Micro hardness values of AA8011-TiC composites.

3.4. Density and Porosity

Figure 9 illustrates the density variations of AA8011-TiC MMCs with different TiC reinforcement percentages (0%, 2%, 4%, 6%, and 8%). Measured density, calculated density, and percentage of porosity were plotted against the weight percentage of particulates. It is demonstrates a positive correlation between TiC reinforcement and density values, indicating that an increase in TiC reinforcement leads to higher density values. It is important to note that the measured densities were slightly lower than the theoretical densities. The discrepancy may be attributed to the limited existence of casting defects in the MMCs, resulting in a small deviation from the expected densities. Nonetheless, the overall trend indicates that the incorporation of TiC reinforcement contributes to an enhancement in the density of the composites. The figure also demonstrates the influence of TiC particle composition on the percent porosity of the AA8011 matrix, with an increasing number of TiC particles leading to an increasing trend in % porosity. Equation (4) [51] was used to apply the Archimedes principle to determine the % porosity, with porosity variations attributed to several factors such as gas entrapment during mixing and shrinkage during solidification.

$$\text{Porosity}(\%) = [1 - \frac{(\text{Theoretical Density} - \text{Measured Density})}{(\text{Theoretical Density})}] \times 100\% \quad (4)$$

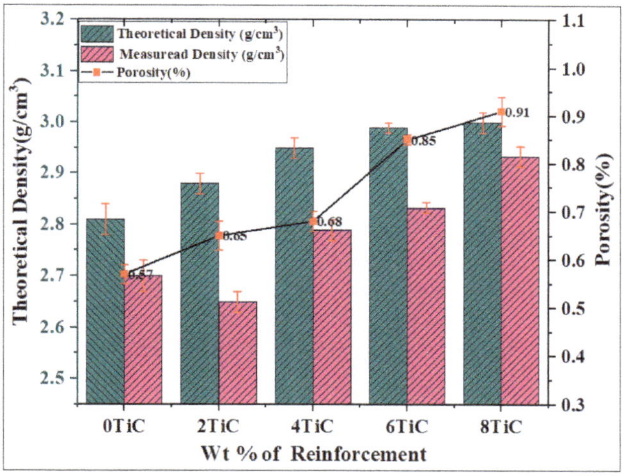

Figure 9. The correlation between experimental density, theoretical density (g/cm^3), and porosity (%) of Al−TiC composites.

3.5. Ultimate Tensile Strength and Elongation

Tensile test specimens were made from as-cast AMCs in accordance with the ASTM E08-8 standard. The stress–strain curves for the composites, depicting the variation observed with different weight fractions of reinforcement, are illustrated in Figure 10. The stress–strain curve analysis provides valuable information about the mechanical response and behaviour of the composites under tensile loading conditions. it is noticeable that all the TiC−reinforced composites have higher tensile strength than the AA8011 alloy [52]. The curves demonstrate the impact of TiC content on the mechanical behaviour of the composites. Furthermore, an increase in TiC content causes a shift of the stress–strain curves towards higher stress levels, indicating an improvement in strength due to the presence of the TiC reinforcement. However, the curves also reveal reduced deformation capacity as the TiC particles hinder the movement of dislocations and restrict plastic deformation.

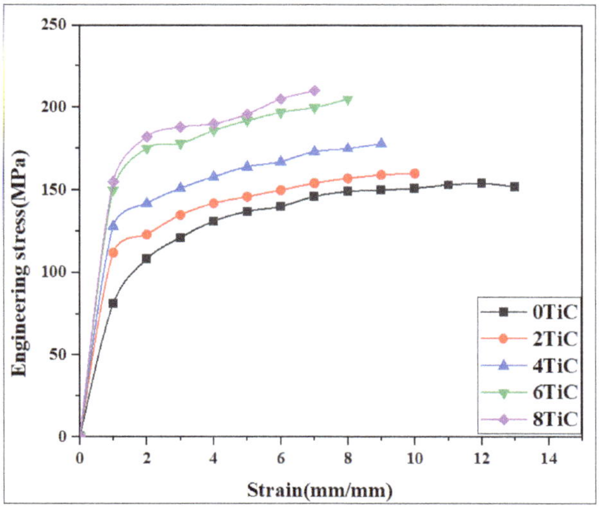

Figure 10. Analysis of stress–strain curves of Al-TiC composites with varying TiC contents.

The tensile strength values of Al-TiC composites are shown in Figure 11. It was noticed that increasing the number of TiC particles in the composite increased its tensile strength. This is owing to the excellent interfacial bonding between the matrix and the reinforcement, which distributes and transfers loads from the matrix to the reinforcement. Among the fabricated composites, it was found that the composite with 8% TiC particles displayed the highest strength. This suggests that incorporating TiC reinforcement improves the overall tensile strength of the composite material. The enhancement is due to the reinforcing effect of the hard TiC particles, which strengthen (Orowan strengthening plus load−bearing effects) the comparatively softer aluminium matrix and improve its tensile resistance. According to the Hall–Petch relationship, the increase in strength can also be attributed to microstructural refinement due to the incorporation of TiC reinforcement [53]. This suggests that as the TiC content increases, there is a compromise in ductility in the composite material [54,55]. The overall strength and ductility response of monolithic and composite samples is displayed in Figure 11.

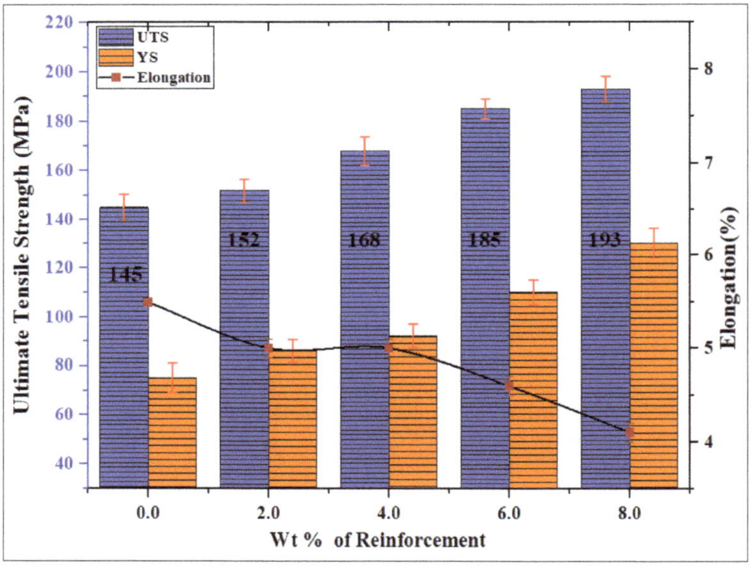

Figure 11. Effect of TiC addition on UTS, yield strength, and % elongation during tensile tests.

Figure 12 illustrates the results of SEM analysis conducted on the fractured surfaces of different specimens. In Figure 12a, the uniformly distributed large voids suggest that the fracture occurred in a ductile manner. This indicates significant plastic deformation and a ductile fracture of the unreinforced alloy. Moving to Figure 12b, the fractured surface of the Al-2 wt.% TiC composite exhibits medium and fine dimples, indicating reasonable plastic deformation. Figure 12c, displays the fractured surface of the Al-4 wt.% TiC composite, showing dimples of varying sizes, clusters of fine dimples, and reinforcement particles within the voids. The presence of deep voids with larger areas can be attributed to the detachment of reinforcement particles and the initiation and propagation of cracks caused by higher stress concentrations near the TiC particulates. Continuing to Figure 12d, the post-failure analysis of the fractured Al-6 wt.% TiC composite revealed a mixed ductile/brittle failure mode characterised by the presence of fine and large dimples, as well as elongated bands [56]. The development of voids within the metal matrix is a primary factor influencing the failure of MMCs. In Figure 12e, for the Al-8 wt.% TiC composites, a greater occurrence of cleavage and the presence of larger dimples indicated a brittle fracture failure.

Finally, Figure 12f, shows the presence of TiC reinforcement on the dendrites' surface, suggesting their presence in interdendritic regions.

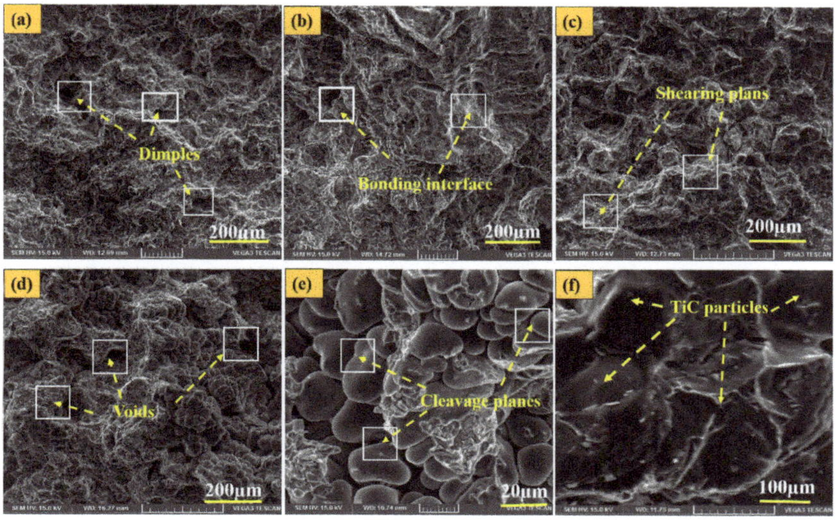

Figure 12. Scanning electron micrographs illustrating fracture morphologies of samples. (**a**) Al-0 wt.% TiC, (**b**) Al-2 wt.% TiC composite, (**c**) Al-4 wt.% TiC composite, (**d**) Al-6 wt.% TiC composite, (**e**) Al-8 wt.% TiC composite, and (**f**) enlarged view highlighting TiC particles covering the dendrites' surface (indicated by dotted yellow arrows).

4. Wear Rate

Figure 13 illustrates the rate of wear of the AA8011 alloy and Al-TiC composite under different loading conditions (10, 20, and 30 N) and different sliding velocities (1, 2, 3, 4, and 5 m/s) and at constant sliding distance 1000 m. It was found that the wear rate of the AA8011 alloy was higher than that of the synthesised Al-TiC composites. On the contrary, the wear rate of the composites displayed a contrasting trend. It demonstrated a linear increase up to a sliding velocity of 1–5 m/s, followed by a gradual decrease at high sliding velocities [22]. The wear rate at 1 m/s sliding velocity is displayed in Figure 13a, depicting the maximum wear rate observed in AA8011 and the minimum wear rate observed in the Al-6 wt.% TiC composite. As the sliding velocity increased, there was a notable rise in the amount of flash temperature generated between the pin surface and the counter disc. This led to the creation of a tribo layer that had an impact on the specific wear rate [57]. At lower sliding speeds (2–3 m/s) and various loads (10, 20, and 30 N), the wear rate of Al-TiC composites exhibited a linear increase, as shown in Figure 13b,c. This behaviour can be attributed to the longer duration of rubbing, leading to increased contact between the surfaces and higher material removal, resulting in elevated wear rates. Conversely, with higher sliding rates (4–5 m/s) and different loads (10, 20, and 30 N), a considerably increased wear rate was noticed, as depicted in Figure 13d,e. Several factors could contribute to this behaviour, including the formation of a protective oxide layer on the composite surface, changes in wear mechanisms, or the presence of surface debris that reduces direct metal-to-metal contact and lowers the wear rate [58]. Among the different volume percentages of TiC composites, the 6 wt.% TiC composite exhibited a relatively low wear rate. This is because the 8 wt.% TiC composites had a higher hardness than the other samples. The wear rate of the 6 wt.% TiC composite increased as the applied load increased while sliding at a speed of 2 m/s. This decline is mostly explained by the increased frictional heat produced at the pin's interaction with the counterface, which led to the oxidation of the pin. As a consequence, there is minimal actual metal−to−metal contact between the pin and the counterface, leading to lower wear rates.

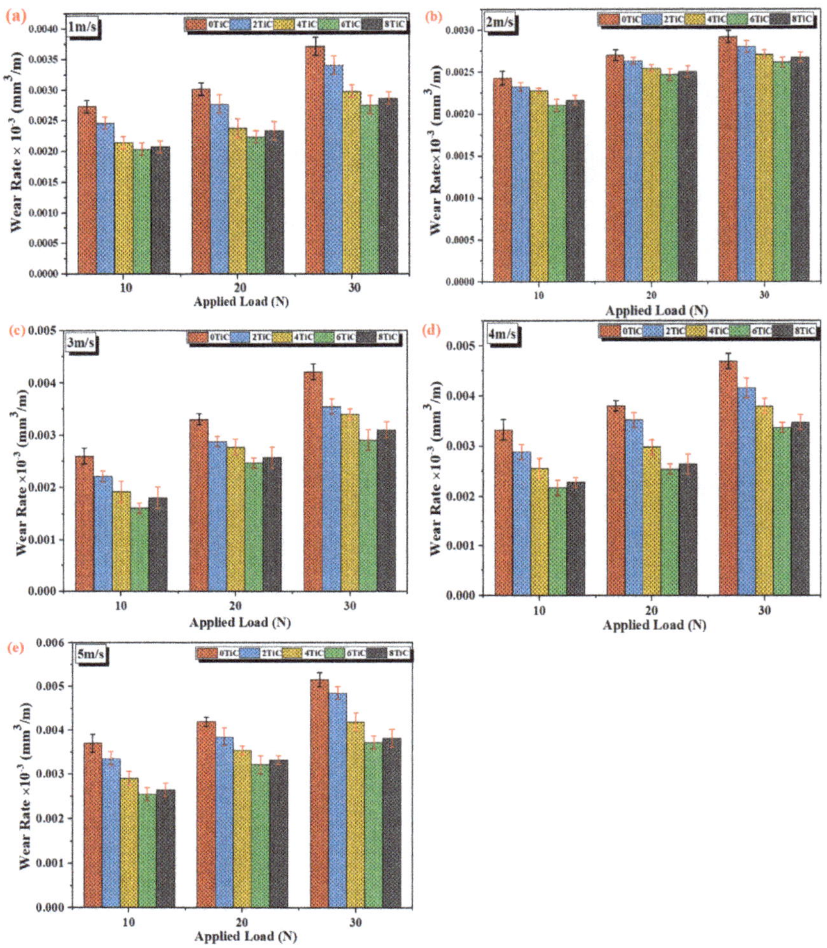

Figure 13. Wear rate as a function of applied load at different sliding velocities for alloy and composite: (**a**) 1 m/s, (**b**) 2 m/s, (**c**) 3 m/s, (**d**) 4 m/s, and (**e**) 5 m/s at a constant sliding distance of 1000 m.

4.1. Coefficient of Friction

Figure 14a–e, depicts the variations in the coefficient of friction (COF) of the AA8011 composite under varying weights of 10, 20, and 30 N at constant sliding distance 1000 m. As illustrated, the coefficient of friction falls as the applied load increases. According to the literature, this behaviour can be related to the production of an oxide layer at lower sliding speeds and a mechanically mixed layer at higher sliding rates of 5 m/s [59]. These layers effectively reduce the actual metallic contact and facilitate easy shearing of the contact interface. During the sliding motion of the pin and counterface, frictional heat is generated at their interface. This heat causes the formation of a mechanically mixed layer, which contains elements from both the pin and counterface materials. This layer is a result of the intense intermixing and deformation of the contacting surfaces. At lower sliding speeds, the formation of an oxide layer also occurs, further reducing the direct metallic contact between the surfaces. In the following sections, we will delve into a detailed discussion of the formation and impact of both the oxide layer and the mechanically mixed layer on the frictional behaviour of the AA8011-TiC composite [60].

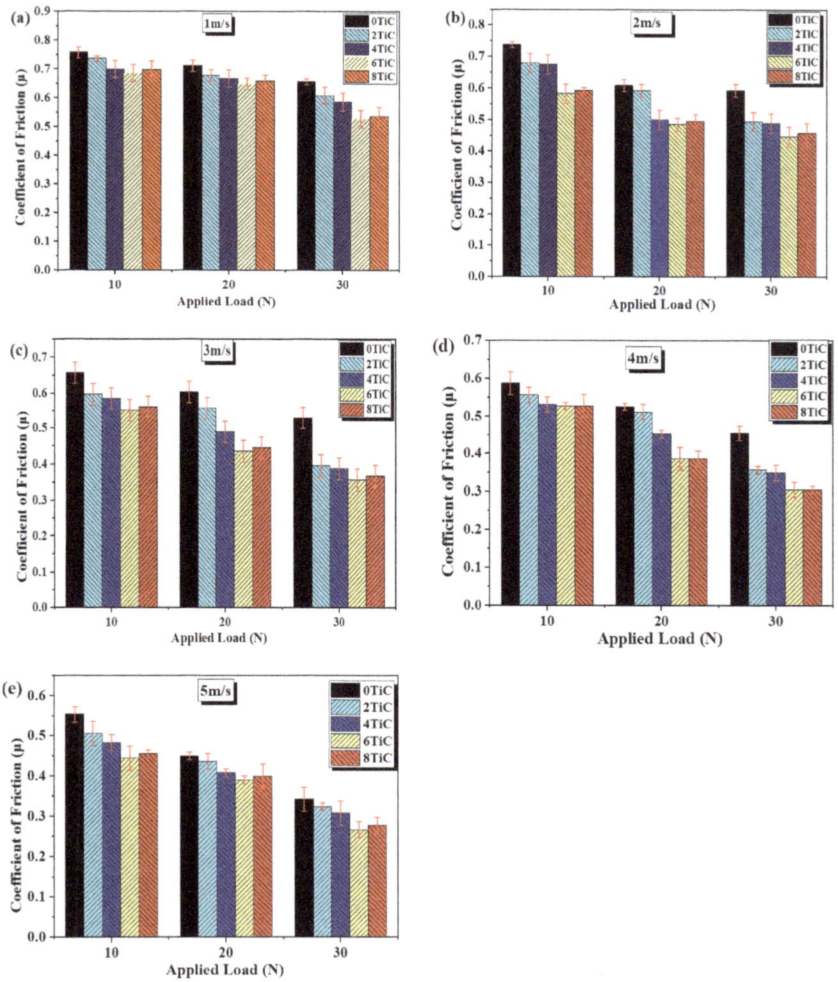

Figure 14. Plotting the coefficient of friction (COF) against the applied load at various sliding velocities for the base alloy and the composites. (**a**) 1 m/s, (**b**) 2 m/s, (**c**) 3 m/s, (**d**) 4 m/s, and (**e**) 5 m/s at a constant sliding distance of 1000 m.

4.2. Worn Pin Surface Analysis

Figure 15 depicts SEM images of worn pin samples investigated under low load conditions (10 N load and 1 m/s). Long grooves parallel to the sliding direction can be seen in all samples, indicating abrasive wear characteristics. However, there are notable distinctions between the base alloy and the TiC-reinforced composite samples. In the base alloy, deeper grooves accompanied by shallow craters are visible (Figure 15a–d) [61]. This effect is caused by the interaction of the soft pin surface and the hard counter surface, which results in low frictional heat and enhanced metallic contact. As a consequence, the hard asperities on the counter surface can penetrate deeper into the base alloy, leading to more pronounced surface damage. This behaviour aligns with the characteristics of delamination wear observed in the base alloy (Figure 15a,b). Delamination wear involves the formation and propagation of subsurface cracks due to repeated sliding under applied loads. As these subsurface cracks propagate and merge, they eventually shear to the surface, resulting in material loss in the form of thin flakes and shallow craters or channels on the worn surface. Conversely, the presence of hard TiC particles within the aluminium alloy matrix adds to the deeper grooves observed in the SEM micrographs of TiC−reinforced composites. The TiC particles serve as abrasive agents, intensifying the abrasive wear mechanism. Additionally, the shallower craters observed in the base alloy samples indicate the occurrence of delamination wear. This wear mechanism involves the nucleation and propagation of subsurface cracks under repeated sliding, eventually leading to surface shearing and the formation of thin flakes and shallow craters or channels [62]. Thus, the wear surface analysis revealed distinct wear mechanisms between the base alloy and the TiC−reinforced composites, with delamination or fatigue wear being dominant in the base alloy and a combination of abrasion and oxidation prevailing in the TiC−reinforced composites [63].

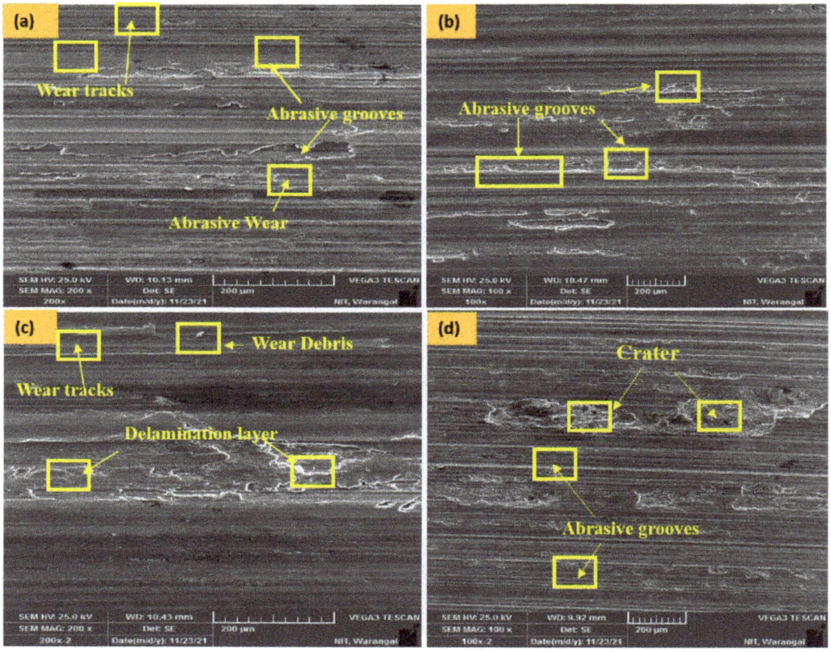

Figure 15. SEM micrographs show characteristics of abrasive wear in the sample examined at 10 N load and 1 m/s speed for (**a**) Al-2 wt.% TiC, (**b**) Al-4 wt.% TiC, (**c**) Al-6 wt.% TiC, and (**d**) Al-8 wt.% TiC.

Figure 16 shows SEM micrographs of the pin samples tested under a load of 20 N and a sliding speed of 3 m/s for all samples. It is noted that the worn pin surfaces of both the base alloys and composites exhibit thin white layers. Through EDS analysis, these spots/layers were identified as oxides, indicating the occurrence of oxidation wear (Figure 16a,b). The oxidation of the pin surface is a consequence of the heat generated during repeated sliding across the counter surface. As time progresses, the oxide debris accumulated in the valleys of the pin surface becomes compacted, leading to the formation of a protective surface oxide layer. The development of this oxide layer on the pin surface plays a crucial role in preventing direct metallic contact with the counter disc, consequently reducing the wear rate [41]. This is supported by the lower wear rates and coefficients of friction observed in the study. The existence of a mechanically mixed layer on the worn surface of the composite is also evident in (Figure 16c,d). This mechanically mixed layer acts as a solid lubricant, promoting increased wear resistance and decreased friction. The presence of oxidation wear and the lubricating properties of the mechanically mixed layer are significant factors contributing to the enhanced wear resistance and reduced friction observed in the composites compared to the base alloys. These findings highlight the beneficial effects of the oxide layer and mechanically mixed layer on the wear response of the composite materials [64].

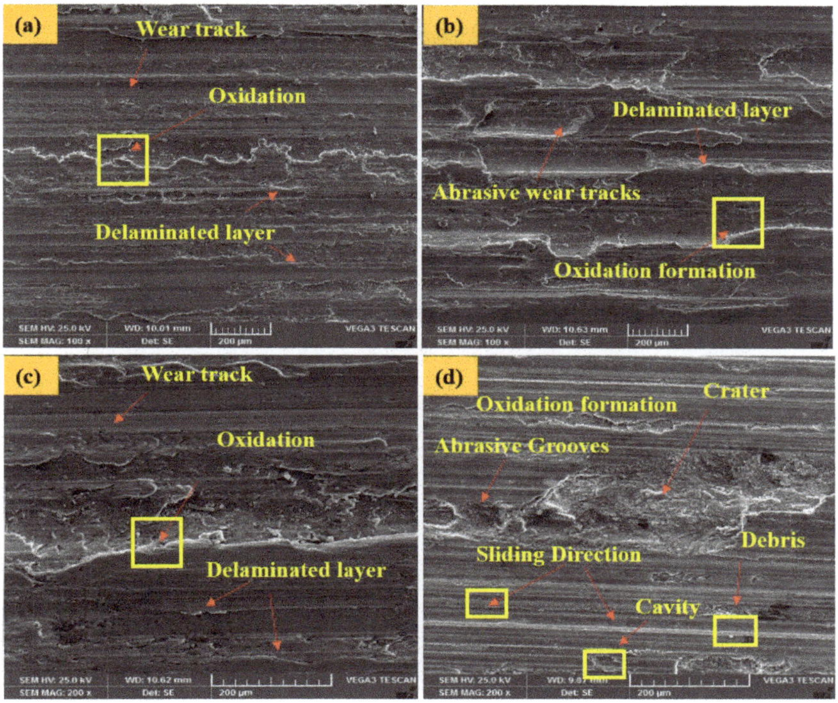

Figure 16. SEM/EDS micrographs show features of oxidation wear and corresponding elemental mapping in the sample examined at 20 N load and 3 m/s speed for (**a**) Al-2 wt.% TiC, (**b**) Al-4 wt.% TiC, (**c**) Al-6 wt.% TiC, and (**d**) Al-8 wt.% TiC.

Figure 17 shows EDS mapping analysis of the worn pin surface of the AA8011-6 wt.% TiC sample, which reveals the presence of iron (Fe) on the surface. This indicates that iron was detected and localised on the worn surface of the composite. The presence of iron may arise from various sources, such as wear debris, interactions between the pin and the counter surface, or iron−rich impurities present in the composite material. The EDS mapping results provide valuable information about the elemental distribution and composition of the worn surface, enhancing comprehension of the wear mechanisms and material responses exhibited by the Al-6 wt.% TiC composite.

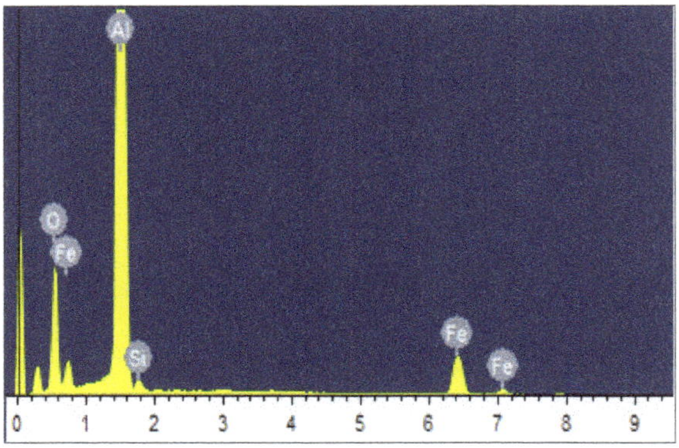

Figure 17. Elemental mapping showing the presence of Fe on the worn surface of the Al−6 wt.% TiC composite sample.

5. Conclusions

The influence of reinforcing TiC into aluminium alloy was studied through investigations, and the following conclusions were drawn.

- A stir casting technique with ultrasonic assistance was employed to fabricate composites with a uniform dispersion of TiC. The manufactured composite exhibits a higher density in comparison to the matrix, primarily due to the high density of TiC particles within its structure.
- TiC particle-reinforced composites were found to possess greater hardness compared to the unreinforced alloy. Among the composites reinforced with TiC particles, those containing 6% exhibited the highest hardness, measured at 61.5%.
- Incorporating TiC particles into the aluminium matrix improves the material's wear resistance. Under high load (30 N) and high-speed (5 m/s) conditions, composites containing 6% TiC particles achieved a maximum wear resistance of 32%.
- Incorporation of TiC particles into the aluminium matrix leads to a decrease in the friction coefficient. Notably, the inclusion of 6 wt.% TiC particles resulted in a significantly low coefficient of friction (0.266) under high load and speed conditions.
- Following a detailed examination of the worn surfaces, it was determined that the base alloy was worn predominantly due to abrasion and delamination. On the other hand, the AA8011 composite samples exhibited oxidation as the dominant wear mechanism.

Author Contributions: Conceptualization, C.B.G., M.B.P. and R.N.R.; methodology, C.B.G., M.B.P., S.I. and M.G.; validation, C.B.G., M.B.P. and R.N.R.; formal analysis, C.B.G., M.B.P. and S.I.; investigation, C.B.G. and M.B.P.; resources, R.N.R.; data curation, C.B.G. and M.B.P.; writing—original draft preparation, C.B.G. and M.B.P.; writing—review and editing, M.B.P., R.N.R., S.I. and M.G.; visualization, C.B.G., M.B.P. and S.I.; supervision, R.N.R. and S.I. All authors have read and agreed to the published version of the manuscript.

Funding: This research received no external funding.

Data Availability Statement: The raw/processed data required to reproduce these findings cannot be shared as the data also forms part of an ongoing study.

Conflicts of Interest: The authors declare no conflict of interest.

References

1. Moorthy, C.V.K.N.S.N.; Kumar, G.N.; Srinivas, V.; Kumar, M.A.; Reddy, K.R.R.M.; Vasundhara, D.N.; Sabri, M.F.M.; Said, S.M. Metallography, Microstructure, and Wear Analysis of AA 6063/TiC Composites for Augmented Dry Sliding Property at Room Temperature. *Metallogr. Microstruct. Anal.* **2020**, *9*, 140–151. [CrossRef]
2. Idrisi, A.H.; Mourad, A.H.I. Conventional Stir Casting versus Ultrasonic Assisted Stir Casting Process: Mechanical and Physical Characteristics of AMCs. *J. Alloys Compd.* **2019**, *805*, 502–508. [CrossRef]

3. Liu, S.; Wang, Y.; Muthuramalingam, T.; Anbuchezhiyan, G. Effect of B_4C and MOS_2 Reinforcement on Micro Structure and Wear Properties of Aluminum Hybrid Composite for Automotive Applications. *Compos. Part B Eng.* **2019**, *176*, 107329. [CrossRef]
4. Mavhungu, S.T.; Akinlabi, E.T.; Onitiri, M.A.; Varachia, F.M. Aluminum Matrix Composites for Industrial Use: Advances and Trends. *Procedia Manuf.* **2017**, *7*, 178–182. [CrossRef]
5. Bulei, C.; Stojanovic, B.; Utu, D. Developments of Discontinuously Reinforced Aluminium Matrix Composites: Solving the Needs for the Matrix. *J. Phys. Conf. Ser.* **2022**, *2212*, 012029. [CrossRef]
6. Kaushik, N.C.; Rao, R.N. Effect of Grit Size on Two Body Abrasive Wear of Al 6082 Hybrid Composites Produced by Stir Casting Method. *Tribol. Int.* **2016**, *102*, 52–60. [CrossRef]
7. Sunar, T.; Özyürek, D. Effect of Al_2O_3Nanoparticles as Reinforcement on the Wear Properties of A356/Al2O3Nanocomposites Produced by Powder Metallurgy. *J. Tribol.* **2022**, *144*, 081701. [CrossRef]
8. Imran Rizwi, M.; Mandal, D.; Show, B.K. Development of Wear Resistant Aluminum Based Nano-Composite Reinforced With Nano-AlN. *J. Tribol.* **2023**, *145*, 021702. [CrossRef]
9. Rao, R.N.; Das, S. Effect of SiC Content and Sliding Speed on the Wear Behaviour of Aluminium Matrix Composites. *Mater. Des.* **2011**, *32*, 1066–1071. [CrossRef]
10. Chandrasekar, P.; Natarajan, S.; Ramkumar, K.R. Influence of Carbide Reinforcements on Accumulative Roll Bonded Al 8011 Composites. *Mater. Manuf. Process.* **2019**, *34*, 889–897. [CrossRef]
11. Sam, M.; Radhika, N. Effect of Heat Treatment on Mechanical and Tribological Properties of Centrifugally Cast Functionally Graded Cu/Al_2O_3 Composite. *J. Tribol.* **2018**, *140*, 021606. [CrossRef]
12. Samal, P.; Vundavilli, P.R.; Meher, A.; Mahapatra, M.M. Influence of TiC on Dry Sliding Wear and Mechanical Properties of in Situ Synthesized AA5052 Metal Matrix Composites. *J. Compos. Mater.* **2019**, *53*, 4323–4336. [CrossRef]
13. Umanath, K.; Palanikumar, K.; Selvamani, S.T. Composites: Part B Analysis of Dry Sliding Wear Behaviour of Al6061/SiC/Al_2O_3 Hybrid Metal Matrix Composites. *Compos. Part B* **2013**, *53*, 159–168. [CrossRef]
14. Moustafa, E.B.; Melaibari, A.; Basha, M. Wear and Microhardness Behaviors of AA7075/SiC-BN Hybrid Nanocomposite Surfaces Fabricated by Friction Stir Processing. *Ceram. Int.* **2020**, *46*, 16938–16943. [CrossRef]
15. Venkatesh, V.S.S.; Rao, R.N.; Patnaik, L. Effect of Spark Plasma Sintering Temperature on Phase Evaluation and Mechanical Behaviour of Cu- 4 Wt% SiC Composite. *Silicon* **2023**. [CrossRef]
16. Pc, E.; Radhakrishnan, G.; Emarose, S. Investigation into Physical, Microstructural and Mechanical Behaviour of Titanium Dioxide Nanoparticulate Reinforced Magnesium Composite. *Mater. Technol.* **2021**, *36*, 575–584. [CrossRef]
17. Fayomi, J.; Popoola, A.P.I.; Popoola, O.M.; Oladijo, O.P.; Fayomi, O.S.I. Results in Physics Tribological and Microstructural Investigation of Hybrid AA8011/ZrB_2-Si_3N_4 Nanomaterials for Service Life Improvement. *Results Phys.* **2019**, *14*, 102469. [CrossRef]
18. Kumaran, S.T.; Uthayakumar, M.; Aravindan, S.; Rajesh, S. Dry Sliding Wear Behavior of SiC and B_4C-Reinforced AA6351 Metal Matrix Composite Produced by Stir Casting Process. *Proc. Inst. Mech. Eng. Part L J. Mater. Des. Appl.* **2016**, *230*, 484–491. [CrossRef]
19. Sivaprasad, K.; Babu, S.P.K.; Natarajan, S.; Narayanasamy, R.; Kumar, B.A.; Dinesh, G. Study on Abrasive and Erosive Wear Behaviour of Al 6063/TiB2 in Situ Composites. *Mater. Sci. Eng. A* **2008**, *498*, 495–500. [CrossRef]
20. Khorasani, M.; Leary, M.; Downing, D.; Rogers, J.; Ghasemi, A.; Gibson, I.; Brudler, S.; Rolfe, B.; Brandt, M.; Bateman, S. Numerical and Experimental Investigations on Manufacturability of Al–Si–10Mg Thin Wall Structures Made by LB-PBF. *Thin-Walled Struct.* **2023**, *188*, 110814. [CrossRef]
21. Ding, Y.; Muñiz-Lerma, J.A.; Trask, M.; Chou, S.; Walker, A.; Brochu, M. Microstructure and Mechanical Property Considerations in Additive Manufacturing of Aluminum Alloys. *MRS Bull.* **2016**, *41*, 745–751. [CrossRef]
22. Sabry, I.; Ghafaar, M.A.; Mourad, A.H.I.; Idrisi, A.H. Stir Casted SiC-Gr/Al6061 Hybrid Composite Tribological and Mechanical Properties. *SN Appl. Sci.* **2020**, *2*, 1–8. [CrossRef]
23. Anand Sekhar, R.; Rakesh Pillai, R.; Mubarak Ali, M.; Shyam Kumar, C. Enhanced Mechanical and Wear Properties of Aluminium-Based Composites Reinforced with a Unique Blend of Granite Particles and Boron Carbide for Sustainable Material Recycling. *J. Alloys Compd.* **2023**, *963*, 171165. [CrossRef]
24. Yadav, M.; Kumaraswamidhas, L.A.; Singh, S.K. Investigation of Solid Particle Erosion Behavior of Al-Al_2O_3 and Al-ZrO_2 Metal Matrix Composites Fabricated through Powder Metallurgy Technique. *Tribol. Int.* **2022**, *172*, 107636. [CrossRef]
25. Bera, T.; Prakash, V.; Acharya, S.K. Erosion Wear Behaviour of A357/Fly Ash Composites. *Proc. Inst. Mech. Eng. Part J J. Eng. Tribol.* **2022**, *236*, 284–291. [CrossRef]
26. Rokni, M.R.; Widener, C.A.; Ozdemir, O.C.; Crawford, G.A. Microstructure and Mechanical Properties of Cold Sprayed 6061 Al in As-Sprayed and Heat Treated Condition. *Surf. Coat. Technol.* **2017**, *309*, 641–650. [CrossRef]
27. Balamurugan, K.; Shanmugam, V.; Palani, G.; Sundarakannan, R.; Sathish, T.; Linul, E.; Khan, S.A.; Asif, M. Effect of TiC/RHA on Solid Particle Erosion of Al6061 Hybrid Composites Fabricated through a 2-Step Ultrasonic-Assisted Stir Casting Process. *J. Mater. Res. Technol.* **2023**, *25*, 4888–4900. [CrossRef]
28. Sakthivelu, S.; Sethusundaram, P.P.; Ravichandran, M.; Meignanamoorthy, M. Experimental Investigation and Analysis of Properties and Dry Sliding Wear Behavior of Al-Fe-Si Alloy Matrix Composites. *Silicon* **2021**, *13*, 1285–1294. [CrossRef]
29. Siddharthan, B.; Bala, Y.V.; Mani, M.B.; Chandresh, A. Experimental Investigation of Tribological Behaviour of Aluminium Alloy Based Metal Matrix Composites. *Bull. Sci. Res.* **2019**, *1*, 10–16. [CrossRef]

30. Fayomi, J.; Popoola, A.P.I.; Popoola, O.M.; Oladijo, O.P.; Fayomi, O.S.I. Understanding the Microstructural Evolution, Mechanical Properties, and Tribological Behavior of AA8011-Reinforced Nano-Si_3N_4 for Automobile Application. *Int. J. Adv. Manuf. Technol.* **2020**, *111*, 53–62. [CrossRef]
31. Keerthivasan, N.; Anandakrishnan, V. Effect of Recycled TiC Addition in the Compaction and Sintering Behavior of Al-Mg-Cu-Zn+r-TiC Powder Metallurgy Composites. *Part. Sci. Technol.* **2022**, *41*, 392–401. [CrossRef]
32. Baskaran, S.; Anandakrishnan, V.; Duraiselvam, M. Investigations on Dry Sliding Wear Behavior of in Situ Casted AA7075–TiC Metal Matrix Composites by Using Taguchi Technique. *J. Mater.* **2014**, *60*, 184–192. [CrossRef]
33. Siddappa, P.N.; Shivakumar, B.P.; Yogesha, K.B.; Mruthunjaya, M.; Girish, D.P. The Effect of TiC Particulate Reinforcement on Dry Sliding Wear Behaviour of Al Based Composites. *Proc. Mater. Today* **2019**, *22*, 2291–2299. [CrossRef]
34. Maleki, K.; Alizadeh, A.; Hajizamani, M. Compressive Strength and Wear Properties of SiC/Al6061 Composites Reinforced with High Contents of SiC Fabricated by Pressure-Assisted Infiltration. *Ceram. Int.* **2021**, *47*, 2406–2413. [CrossRef]
35. Kumar, R.; Dhiman, S. A Study of Sliding Wear Behaviors of Al-7075 Alloy and Al-7075 Hybrid Composite by Response Surface Methodology Analysis. *Mater. Des.* **2013**, *50*, 351–359. [CrossRef]
36. Rao, R.N.; Das, S. Effect of Applied Pressure on the Tribological Behaviour of SiCp Reinforced AA2024 Alloy. *Tribol. Int.* **2011**, *44*, 454–462. [CrossRef]
37. Kumar, H.G.P.; Xavior, M.A. Assessment of Mechanical and Tribological Properties of Al 2024- SiC—Graphene Hybrid Composites. *Procedia Eng.* **2017**, *174*, 992–999. [CrossRef]
38. Prasad Reddy, A.; Vamsi Krishna, P.; Rao, R.N. Two-Body Abrasive Wear Behaviour of AA6061-2SiC-2Gr Hybrid Nanocomposite Fabricated through Ultrasonically Assisted Stir Casting. *J. Compos. Mater.* **2019**, *53*, 2165–2180. [CrossRef]
39. Prasad Reddy, A.; Vamsi Krishna, P.; Rao, R.N. Tribological Behaviour of Al6061–2SiC-XGr Hybrid Metal Matrix Nanocomposites Fabricated through Ultrasonically Assisted Stir Casting Technique. *Silicon* **2019**, *11*, 2853–2871. [CrossRef]
40. Liu, Z.; Han, Q.; Li, J. Ultrasound Assisted in Situ Technique for the Synthesis of Particulate Reinforced Aluminum Matrix Composites. *Compos. Part B Eng.* **2011**, *42*, 2080–2084. [CrossRef]
41. Jeyasimman, D.; Narayanasamy, R.; Ponalagusamy, R.; Anandakrishnan, V.; Kamaraj, M. The Effects of Various Reinforcements on Dry Sliding Wear Behaviour of AA 6061 Nanocomposites. *Mater. Des.* **2014**, *64*, 783–793. [CrossRef]
42. Pasha, M.B.; Sharma, R.C.; Rao, R.N.; Ismail, S.; Gupta, M. Sliding Wear Characteristics of Mg/Si_3N_4 Nanocomposites at Room and Elevated Temperatures. *Mater. Lett.* **2022**, *329*, 133186. [CrossRef]
43. Rao, R.N.; Das, S. Effect of Matrix Alloy and Influence of SiC Particle on the Sliding Wear Characteristics of Aluminium Alloy Composites. *Mater. Des.* **2010**, *31*, 1200–1207. [CrossRef]
44. Bharat, N.; Bose, P.S.C. Influence of Nano-TiO_2 Particles on the Microstructure, Mechanical and Wear Behaviour of AA7178 Alloy Matrix Fabricated by Stir Casting Technique. *Proc. Inst. Mech. Eng. Part L J. Mater. Des. Appl.* **2022**, *237*, 753–766. [CrossRef]
45. Li, Y.; Ramesh, K.T.; Chin, E.S.C. Comparison of the Plastic Deformation and Failure of A359/SiC and 6061-T6/Al_2O_3 Metal Matrix Composites under Dynamic Tension. *Mater. Sci. Eng. A* **2004**, *371*, 359–370. [CrossRef]
46. Lijay, K.J.; Selvam, J.D.R.; Dinaharan, I.; Vijay, S.J. Microstructure and Mechanical Properties Characterization of AA6061/TiC Aluminum Matrix Composites Synthesized by in Situ Reaction of Silicon Carbide and Potassium Fluotitanate. *Trans. Nonferrous Met. Soc. China* **2016**, *26*, 1791–1800. [CrossRef]
47. Thangarasu, A.; Murugan, N.; Dinaharan, I. Production and Wear Characterization of AA6082-TiC Surface Composites by Friction Stir Processing. *Procedia Eng.* **2014**, *97*, 590–597. [CrossRef]
48. Wang, F.; Liu, H.; Yang, B. Effect of In-Situ TiC Particulate on the Wear Resistance of Spray-Deposited 7075 Al Matrix Composite. *Mater. Charact.* **2005**, *54*, 446–450. [CrossRef]
49. Salur, E.; Acarer, M.; Şavkliyildiz, İ. Improving Mechanical Properties of Nano-Sized TiC Particle Reinforced AA7075 Al Alloy Composites Produced by Ball Milling and Hot Pressing. *Mater. Today Commun.* **2021**, *27*, 102202. [CrossRef]
50. Ramakoteswara Rao, V.; Ramanaiah, N.; Sarcar, M.M.M. Dry Sliding Wear Behavior of TiC -AA7075 Metal Matrix Composites. *Int. J. Appl. Sci. Eng.* **2016**, *14*, 27–37. [CrossRef]
51. Pasha, M.B.; Rao, R.N.; Ismail, S.; Özcan, M.; Prasad, P.S.; Gupta, M. Assessing Mg/Si_3N_4 Biodegradable Nanocomposites for Osteosynthesis Implants with a Focus on Microstructural, Mechanical, in Vitro Corrosion and Bioactivity Aspects. *J. Mater. Res. Technol.* **2022**, *19*, 3803–3817. [CrossRef]
52. Yang, Y.; Liu, Z.; Jiang, R.; Li, R.; Li, X. Microstructural Evolution and Mechanical Properties of the AA2219/TiC Nanocomposite Manufactured by Ultrasonic Solidification. *J. Alloys Compd.* **2019**, *811*, 151991. [CrossRef]
53. Xi, H.; Xiao, W.; Li, H.; Fu, Y.; Yi, G.; Qie, J.; Ma, X. Effects of Submicron-Sized TiC Particles on the Microstructure Modification and Mechanical Properties of Al-Si-Mg Alloy. *J. Alloys Compd.* **2023**, *968*, 171963. [CrossRef]
54. Ravi Kumar, K.; Kiran, K.; Sreebalaji, V.S. Micro Structural Characteristics and Mechanical Behaviour of Aluminium Matrix Composites Reinforced with Titanium Carbide. *J. Alloys Compd.* **2017**, *723*, 795–801. [CrossRef]
55. Li, B.; Chen, X.; Jiang, M.; Ma, S.; Wang, Z.; Lei, Z.; Chen, Y. Microstructure and Mechanical Properties of TiC Nanoparticles Reinforced 7075 Aluminium Alloy Fabricated by Oscillating Laser-Arc Hybrid Additive Manufacturing. *Sci. Technol. Weld. Join.* **2023**, *28*, 495–505. [CrossRef]
56. Munivenkatappan, M.S.B.; Shanmugam, S.; Veeramani, A. Synthesis and Characterization of In-Situ AA8011-TiB_2 Composites Produced by Flux Assisted Synthesis. *Ann. Chim. Sci. Mater.* **2020**, *44*, 333–338. [CrossRef]

57. Kumar, A.; Gautam, R.K.; Tyagi, R. Dry Sliding Wear Characteristics of in Situ Synthesized Al-TiC Composites. *Compos. Interfaces* **2016**, *23*, 469–480. [CrossRef]
58. Archard, J.F. Contact and Rubbing of Flat Surfaces. *J. Appl. Phys.* **1953**, *24*, 981–988. [CrossRef]
59. Baradeswaran, A.; Elaya Perumal, A. Influence of B_4C on the Tribological and Mechanical Properties of Al 7075-B4C Composites. *Compos. Part B Eng.* **2013**, *54*, 146–152. [CrossRef]
60. Ul Haq, M.I.; Anand, A. Dry Sliding Friction and Wear Behavior of AA7075-Si_3N_4 Composite. *Silicon* **2018**, *10*, 1819–1829. [CrossRef]
61. Rao, R.N.; Das, S. Effect of Sliding Distance on the Wear and Friction Behavior of as Cast and Heat-Treated Al-SiCp Composites. *Mater. Des.* **2011**, *32*, 3051–3058. [CrossRef]
62. Ul Haq, M.I.; Anand, A. Friction and Wear Behavior of AA 7075- Si_3N_4 Composites Under Dry Conditions: Effect of Sliding Speed. *Silicon* **2019**, *11*, 1047–1053. [CrossRef]
63. Khan, M.M.; Nisar, M. Effect of in Situ TiC Reinforcement and Applied Load on the High-Stress Abrasive Wear Behaviour of Zinc–Aluminum Alloy. *Wear* **2022**, *488–489*, 204082. [CrossRef]
64. Vinoth, B.; Alagarsamy, S.V.; Meignanamoorthy, M.; Ravichandran, M. Prediction of Tribological Performance of AA8011/Wt.%ZrO_2 Based Composites Fabricated by Stir Casting Route. *Proc. Inst. Mech. Eng. Part E J. Process Mech. Eng.* **2022**, *236*, 2420–2433. [CrossRef]

crystals

MDPI

Article

On the Efficient Particle Dispersion and Transfer in the Fabrication of SiC-Particle-Reinforced Aluminum Matrix Composite

Andong Du [1,2], Lucia Lattanzi [3], Anders E. W. Jarfors [2,3,*], Jinchuan Zheng [2], Kaikun Wang [1,*] and Gegang Yu [2]

[1] Department of Materials Processing and Control Engineering, School of Materials Science and Engineering, University of Science and Technology Beijing, Xueyuan Road 30, Haidian District, Beijing 100083, China; b20170186@xs.ustb.edu.cn

[2] China Academy of Machinery Sciences and Technology (Jiangle) Institute of Semi-Solid Metal Technology, Huancheng East Road 22, Jiangle County, Sanming 353300, China; zhengjc@cam.com.cn (J.Z.); yugg@cam.com.cn (G.Y.)

[3] Department of Materials and Manufacturing, School of Engineering, Jönköping University, P.O. Box 1026, 551 11 Jönköping, Sweden; lucia.lattanzi@ju.se

* Correspondence: anders.jarfors@ju.se (A.E.W.J.); kkwang@ustb.edu.cn (K.W.); Tel.: +46-36101651 (A.E.W.J.)

Abstract: Lightweight SiC-particle-reinforced aluminum composites have the potential to replace cast iron in brake discs, especially for electric vehicles. This study investigates the effect of SiC particle size and matrix alloy composition on the resulting transfer efficiency and particle distribution. The performance of a specially designed stirring head was studied using a water model, and the stirring head conditions were assessed to understand the particle transfer and dispersion mechanisms in the molten aluminum. The standard practice of thermal pre-treatment promotes the wetting of the reinforcing particles and commonly causes clustering before the addition to the melt. This early clustering affects the transfer efficiency and particle dispersion, where their interaction with the melt top-surface oxide skin plays an important role. In addition, the transfer efficiency was linked to the particle size and the chemical composition of the matrix alloy. Smaller particles aggravated the degree of clustering, and the addition of rare earth elements as alloying elements in the matrix alloy affected the particle dispersion. The stirring parameters should be selected to ensure cluster disruption when the carbides are added to the melt.

Keywords: metal matrix composites; microstructure; casting; simulation

1. Introduction

Silicon carbide (SiC) presents more than two hundred crystalline forms, many of these are rare in nature, but they can be artificially synthesized. Thanks to its interesting thermophysical and electrical properties, SiC has been highly used in thermal isolation materials, microelectronics, optoelectronics, and nuclear applications [1–4]. SiC particles are especially interesting for producing metal matrix composites; lightweight SiC particle-reinforced aluminum (Al) composites have attracted attention in recent years for their excellent durability compared with cast iron in automotive brake discs [5]. Nowadays, legislation and requirements rapidly evolve towards stricter gas and particle emission limits, like the recent Euro 7 standard [6]. Thus, automotive and train manufacturers are becoming increasingly interested in developing lightweight components of Al alloys and their composites [7,8]. A lightweight brake system reduces the unsprung mass, improves road contact and driver experience, and, more importantly, reduces particulate matter and greenhouse gas emissions [9]. The brake discs must be durable, withstand wear, and provide friction to function correctly. Moreover, the braking system, as such, must be thermally stable [10]. The SiC particle (SiCp)-reinforced Al metal matrix composite (MMC), thanks to its high modulus and strength, high specific weight ratio, and wear resistance, is suitable for brake disc application.

Several methods are available to fabricate SiCp-MMC components, from powder metallurgy to spray deposition, from pressure infiltration to stir casting. From a commercial perspective, ease of fabrication and low cost are critical factors [11]. The stir-casting method satisfies these two factors, but there are still challenges in other production steps like machining and drilling. During the casting operations, the ambient atmosphere/air may be entrained in the molten metal during the stirring process [12]. Furthermore, in the Duralcan™ stir casting process, there are still limitations in interfacial bonding [13], and the micrometer-sized particles tend to agglomerate. Altogether, the particle dispersion in the molten Al is complex and challenging [14]. The wettability between SiCp and molten Al also affects the dispersion of particles. Bao et al. [12] employed the sessile drop experiment to test the effect of temperature on the contact angle between the molten Al and SiCps. The authors reported that a heat treatment at 1100 °C for 135 min reduced the contact angle from 136° to 55°. According to the Young–Dupre equation [15], there are three ways to improve the wettability of SiC with molten Al: by increasing the surface energy of SiC to create a coating of SiO_2 film; by decreasing the surface tension of the molten Al by adding the reactive elements magnesium (Mg), titanium (Ti), and lithium (Li), which could react with the oxide Al film; and, as such, by decreasing the interfacial energy between the SiC and molten Al. Urena et al. [16] proposed that a SiO_2 layer is generated on the SiCp surface after oxidation treatment of the particles; the SiO_2 layer reduced the interface free energy, promoted wetting, and hindered the formation of the detrimental Al_4C_3. The SiC particle thermal pre-treatment and adding alloying elements to the matrix alloy are viable methods to hinder the interface chemical reaction.

Priyan et al. [17] investigated the effect of ultrasonic treatment on the dispersion of 4 wt.% of SiC particles in stir casting, and the mean inter-particle distance was the parameter used to characterize the particle dispersion property. Priyan et al. [17] reported that the mean inter-particle distance increased after the ultrasonic treatment, indicating a de-clustering effect. Fan et al. [18] suggested a new method to uniformly disperse the SiCps using a rotor and stator unit to generate high-intensity shearing to condition the melt for further processing. This method was developed for magnesium-based composites reinforced with SiC particles, which were successfully uniformly dispersed. Yang et al. [19] used a rotor and stator intensive-shearing stirring head to fabricate an Al-7Si-based composite reinforced with 10 vol.% of SiCps. Uniform particle dispersion was achieved, and the material's tensile strength increased by intensive shearing, which successfully broke the SiCp clusters. The stirring head hindered the gas entrapment and reduced the casting defects.

From the above discussion, clustering is a challenge, and it is essential to understanding cluster behavior better. Capes [20] investigated the correlation between agglomeration strength and the size of sand particles and reported that the strength factor increased as the sand particle size decreased, indicating that the particle size decreased as the bonding point per area increased. Wilks [21] studied the effect of homogeneous SiC reinforcement on the chevron notch fracture toughness and reported that the toughness increased two-fold after breaking the SiCp clusters. Not only could the homogeneous SiCp dispersion improve the properties of the composite, but also the particle transfer efficiency. However, due to the limits of the stirring process, not all the SiC particles are transferred into the melt. Particle transfer efficiency is a critical factor that is as important as particle dispersion since it directly affects the number of particles that contribute to improving the composite's properties.

In previous works, La and Ce elements were added to the matrix alloy of Al-based MMCs to increase their mechanical resistance at high temperatures [22–24] for their use in train brake discs. The same composite materials were investigated in the present study from a process point of view to evaluate particle dispersion and transfer efficiency. This study investigates what hinders a high transfer efficiency of SiC particles to an Al melt and a good dispersion of reinforcement with a low level of clustering. The state-of-the-art and best practices for the thermal pre-treatment of stirring conditions were taken as a starting point to have a direct link to commercial processing. This study focuses on the effect of two factors: the chemical composition of the matrix alloy and particle size and size distribution. The goal is to describe the role played by the SiC particles' characteristics and the alloying elements in the matrix alloy in influencing the transfer and agglomeration of carbides. The results help select the best process parameters to obtain a homogeneously dispersed composite.

2. Materials and Methods

The current paper investigates the introduction and dispersion of SiC particles into an Al alloy melt. The workflow consists of the following steps.

1. Preparation of the composite materials to study the effects of SiCp size and size mix as well as the addition of rare earth elements that have an affinity to oxygen and may change the melt's wetting behavior.
2. Water modeling with torque and rate of rotation assessment to estimate the shear forces in the pumping head to assess the possibility of breaking SiCp clusters and alumina bifilms that appear when introducing the SiC particles through the melt top surface.
3. Simulation work to understand the flow speeds and general dispersion of the particles in the crucible and to assess the absence of stagnant zones and pseudo cavern formation.
4. Metallographic analysis to assess the transfer efficiency and level of agglomeration and clustering. The resulting data is used together with steps 2 and 3 to explain the origin of losses in particle transfer, particle dispersion, and the agglomeration/deagglomeration effect.

2.1. Material Production and Characterization

2.1.1. Material Production

Five different composite materials were produced by squeeze casting with different SiC particle sizes and matrices. The targeted addition of SiC particles was 20 wt.%, as Jarfors et al. [25] proposed that 20 wt.% of SiC particles could provide enough friction force for the brake disc application. The matrix alloy formulation was based on previous work [22] and the currently available commercial application for automotives [5,26]. The chemical composition of the composite materials, listed in Table 1, were evaluated with optical emission spectroscopy. The Si values include the Si of the matrix alloy and the Si in the SiC particles. The material production process was described in recent work [27].

Table 1. Chemical composition of composite materials [wt.%], including the SiC particles [1].

Material	SiCp Size [μm]	Si	Fe	Cu	Mn	Mg	Ni	Ti	Ce	La
C0_23	23	15.8	0.7	0.0	0.0	0.7	0.0	0.1	-	-
C0_50	50	21.2	0.5	0.0	0.0	0.7	0.0	0.1	-	-
C0_10	10	10.9	0.4	0.1	0.0	0.6	0.1	0.1	-	-
C0_mix	10 + 23 + 50 [2]	17.4	0.2	0.0	0.0	0.8	0.0	0.1	-	-
C1_23	23	19.8	0.5	1.9	0.9	0.8	1.9	0.4	0.6	0.7

[1] The carbon was not included in the analysis due to equipment limitations. [2] The different sizes were added at equal mass ratios in the material C0_mix.

A newly designed rotor–stator stirring device made of SiC ceramic was used for the stirring process, and the design was modified from the original version of the device developed by Yang et al. [19]. The device's schematic was also described in previous work [27]; the schematic structure and dimensions of the rotor–stator device are shown in Figure 1. The rotor–stator device was made of silicon carbide, and the crucible was made of clay-bonded graphite. The SiC particles are poured on the molten metal's surface and then sucked into the melt by the action of the rotor–stator unit. No inert gas protection was used, and the rotor speed was limited to 500 rpm to avoid gas entrainment/cavitation effects and keep the melt surface quiescent while stirring for one hour. The SiC particles were pre-treated at 1000 °C under air atmosphere conditions for 1 h to form a thin SiO_2

layer on the SiCp particles [16]. Squeeze casting with 100 MPa of pressure was used to minimize the casting defects, and the mold dimensions' details were shown in previous work [22]. Each casting experiment produced 4 kg of composite and 30 samples. Randomly, one sample was cut and polished for metallographic observations.

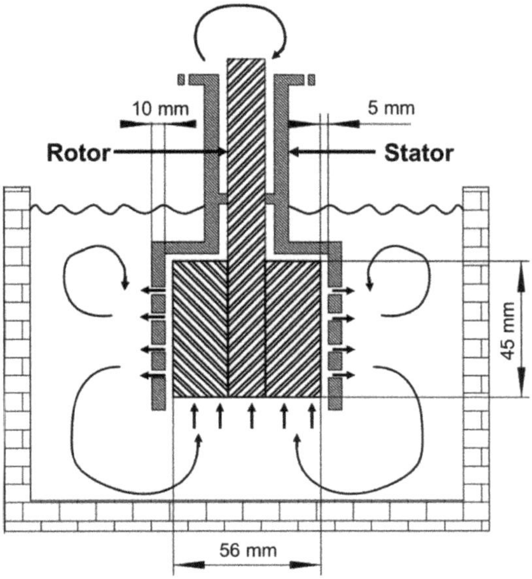

Figure 1. The schematic structure and dimensions of the rotor–stator device.

2.1.2. Microstructural Characterization

The samples were prepared for metallographic observation by combining the grinding steps for a hard material and the polishing steps for a soft material. In this case, the SiC particles represent the hard material, and the Al-Si matrix is the soft material. The sequence of grinding and polishing discs (Struers, Ballerup, Denmark) was Piano 220 with water, Largo with 9 µm diamond suspension, Dac with 3 µm diamond suspension, and Dur with 1 µm diamond suspension. Metallographic observations were performed on an optical microscope, the Olympus DSX1000 (Olympus Corporation, Shinjuku, Japan). The SEM observations were performed on the scanning electron microscope JSM-7001f (JEOL, Akishima, Japan).

The SiC particles and particle clusters were analyzed by the MIPAR Image Analysis software (version 4.0.0) on panorama micrographs that covered the entire surface of the samples. The area fraction and diameter of the SiCp clusters were used to evaluate the particles' dispersion. The particle transfer efficiency is defined in Equation (1) as the ratio between the amount of SiCp evaluated through the image analysis (IA) and the targeted content of SiCps.

$$\text{SiCp transfer efficiency} = \frac{\text{SiCwt.\% achieved}}{\text{SiCwt.\% targeted}} \quad (1)$$

The nearest neighbor distance was calculated based on the centroid coordinates of the SiC particles and the 'knnsearch' function in MATLAB R2022b. The average of the nearest neighbor values was used to evaluate the distribution of the carbides.

2.2. *Modeling*

2.2.1. Water Modeling

A glass tank full of water was used for the water modeling, and the rotor and stator were fabricated by 3D printing (Figure 2). The servo motor (Delta, ASDA-B2) was used for the rotor motion, and the servo controller (Delta, ASDA-B2) was used to control velocity and collect torque and power consumption values.

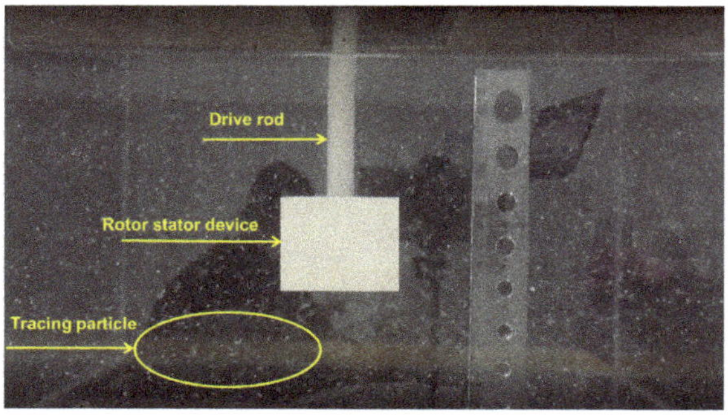

Figure 2. Experimental setup of the water modeling, showing the drive rod, rotor–stator device, and tracing particles for flow visualization. The vertical ruler serves as a scale bar.

The tracing particles are lightweight white wax-based particles (Amer Technology Co. Ltd., Guangzhou, China) that are free to move in water to visualize the flow and verify if the rotor–stator device could generate a strong enough flow to disperse the particles in the water homogeneously.

The structure of the rotor–stator device and the experiment parameters in the water modeling experiment were similar to those used in the Al melt stirring experiment. The steady-state velocity gradient, $\partial u/\partial z$, depends on the geometry only, so using the same structure rotor–stator unit in the water modeling experiment resulted in similar experimental parameters to those in the Al melt stirring experiment.

For this reason, using water modeling to evaluate the torque and thus the velocity gradient, $\partial u/\partial z$, in the Al melt stirring experiment was possible. In the measurements from the water modeling experiment, the torque, T_{water}, is measured to calculate the shear force at the rotor according to Equation (2).

$$\tau_{\text{water}} = \frac{T_{water}}{A \times r} \quad (2)$$

In Equation (2), A is the blade tip area (8.1 10^{-4} m^2), r is the radius of the rotor (0.028 m), and τ_{water} is the shear force generated by the rotor–stator device in the water modeling experiment. This calculation gives an effective measure of the speed gradient in the rotor–stator pumping arrangement, and it is related to the viscosity as in Equation (3). Using τ_{water} [Pa] and the water viscosity, μ_{water} (1 mPa s) [28], in Equation (3), it is possible to obtain $\partial u/\partial z$ of in the water modeling experiment.

$$\frac{\partial u}{\partial z} = \frac{\tau_{water}}{\mu_{water}} \quad (3)$$

The assumption is that the system is under a steady state; thus, the speed gradient is only geometry-dependent and, therefore, can be used for both the water and the composite slurry. Equation (4) clarifies the relationship:

$$\tau_{MMC} = \mu_{MMC}\frac{\partial u}{\partial z} \equiv \mu_{MMC}\frac{\tau_{water}}{\mu_{water}}, \quad (4)$$

where τ_{MMC} is the shear stress generated by composite slurry stirring and μ_{MMC} is the viscosity of the composite slurry.

2.2.2. Flow-3D Simulation

The commercial Flow-3D software (version 11.2) simulated the SiC particle stirring and dispersion process, and the structure size of the rotor and stator was the same as for the casting experiment. The AlSi10Mg alloy was chosen for this simulation. The parameters of velocity and temperature were fixed to 500 rpm and 700 °C, respectively. A cartesian grid mesh with a cell size of 1.5 mm and a time step of 0.001 s was used. The applied models are three-dimensional with gravity and a non-inertial reference frame; a general moving object model was used for the particles with a random distribution; and the turbulence model was the renormalized group (RNG) model. The boundary conditions were no wall shear stress or pressure on the top surface. The tracers representing the SiC particles in the simulation were described as spheres with a diameter of 23 μm and a density of 3.2 g/cm^3, and the volume fraction was 1% for computational efficiency as their effect on viscosity was regarded as limited [29]. The simulation described the flow pattern of the Al melt during the stirring process.

3. Results and Discussion

The results and discussion will start with the transfer efficiency, followed by the deagglomeration and dispersion of the particles into the melt.

3.1. Particle Transfer Efficiency

Table 2 shows the SiCp transfer efficiency, and data were obtained through the metallographic observation of fifteen images.

Table 2. Particle image analysis and evaluation of the transfer efficiency. The quantitative image analysis was performed on panorama images that covered the entire surface of the samples.

MMC	C0_23	C0_50	C0_10	C0_Mix	C1_23
Average particle size [μm]	14 ± 0.9	32 ± 1.4	12 ± 1.8	19 ± 1.5	15 ± 1.1
SiCp wt.% targeted fraction	20%	20%	20%	20%	20%
SiCp wt.% fraction	8.7%	15.2%	4.8%	8.5%	12.4%
SiCp transfer efficiency	43.6%	75.8%	24.1%	42.2%	61.9%

The material C0_50 presented a 75.8% transfer efficiency, which was the highest value. The material C0_10 presented a transfer efficiency of 24.1%, which was the lowest value, while the mixed-size particles transferred into the melt with a 42.2% transfer efficiency. The materials with the same size of SiCp but with a different chemical composition of the matrix alloy, materials C0_23 and C1_23, presented transfer efficiencies of 43.6% and 61.9%, respectively. Figure 3 represents the transfer efficiency in relation to the particle size and shows a positive correlation between the two factors.

The material C1_23 was not included in the fitting due to its different matrix composition. The material C1_23 shows a higher transfer efficiency compared with the material C0_23, even though they have the same particle size. This suggests that the La/Ce addition did affect the transfer efficiency.

Figure 4 shows the structure and morphology of the SiC particles after the thermal pre-treatment and before their addition to the melt. After the thermal pre-treatment, the SiC particles tended to agglomerate due to a slight sintering tendency. Figure 4a shows the particles with a particle size of 23 μm, and the agglomeration is visible. Figure 4b shows the particles with a size of 50 μm, the largest particles used in this research, and an agglomeration of particles is not observed. Figure 4c shows the particles with a size of 10 μm, the smallest particles used in the present study, and they show clustering similar to the 23 μm-sized particles. Figure 4d shows the particles with mixed sizes, with a ratio of 1:1:1 = 23 µm:50 µm:10 µm. The small particles with a size of 10 µm adhere to other particles, forming satellite particles on the surfaces of the larger particles with sizes of 23 µm and 50 µm. Figure 4 shows that the mixed-sized particles appear clustered before their addition to the melt, which might affect the transfer efficiency. The material C0_50 shows the highest transfer efficiency, and the particles were not agglomerated before adding the molten metal.

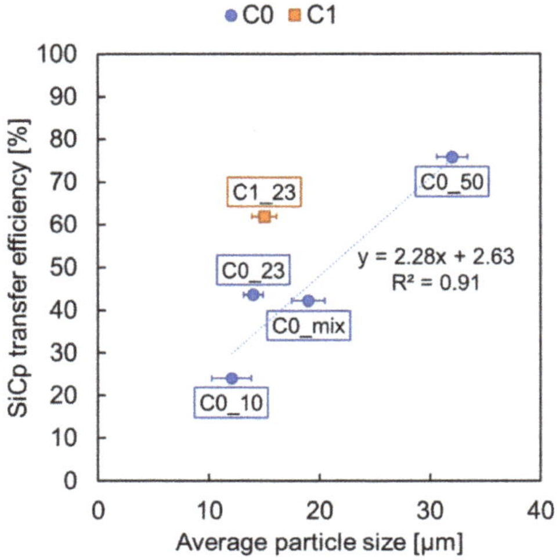

Figure 3. A diagram of particle transfer efficiency vs. average particle size.

Figure 4. Appearance of SiC particles after the thermal pre-treatment: (**a**) 23 µm; (**b**) 50 µm; (**c**) 10 µm; (**d**) mix of the three dimensions.

The SiCp particles added in a clustered state mean that each cluster requires the top surface of the molten metal to be broken. The presence of aluminum oxide layers characterizes the surface of the molten aluminum, also referred to as bifilms, according to research by Campbell [30]. These bifilms can interact with the SiCp clusters and wrap the particles, making their dispersion in the melt more difficult. The air sandwiched between the SiC particles of the cluster decreased the density of the clusters, making them highly buoyant. During the addition of SiCps into the melts, the stirring force is a critical factor in transferring the particles into the melts to counteract floatation and settling. The aluminum oxide film encapsulating the SiCp clusters must be fractured to allow melt infiltration. This phenomenon would then decrease the transfer efficiency of the particles. Once the aluminum oxide film is broken, wetting is possible.

Compared to material C0_23, the transfer efficiency of the material C1_23 was 42.5% higher, which indicated that the addition of La and Ce did affect the transfer efficiency. Liu et al. [31] indicated that La and Ce react with Al_2O_3, forming $(La,Ce)_2O_3$, which could reduce the surface energy and promote the wettability of SiCps. According to the Young–Dupre equation [15], reducing the surface energy could decrease the contact angle and promote wettability between the SiC and Al melts. Lloyd [13] reviewed that Mg reacted with Al_2O_3, forming the spinel crystal $MgAl_2O_4$, which increases the wettability of SiC in the molten Al alloy. For this reason, it is interesting to compare the well-known behavior of Mg with La and Ce to better understand their possible contribution in terms of wetting. Table 3 shows the standard Gibbs energy change of oxidation reactions from the Ellingham diagrams [32]; the larger the negative value of the standard Gibbs energy change of the formation ΔG°_{f} [kJ/(mol O_2)], the more stable the oxide is. According to the data presented in Table 3 by Hasegawa [32], La has a larger negative value than Mg, which means the affinity of La to O is higher than that of Mg to O, and oxygen (O) is more likely to react with La than with Mg for spinel formation. Based on these values, one can conclude that La and Ce have a behavior similar to Mg in promoting the transfer efficiency of SiC particles.

Table 3. The standard Gibbs energy change of oxidation reactions by Hasegawa [32].

Element	La	Ce	Mg	Al	Ti	Si	Mn	Fe	Cu
ΔG°_{f} [kJ/(mol O_2)]	−1055	−880	−992	−900	−845	−728	−624	−400	−200

3.2. *Deagglomeration of Clusters*

3.2.1. SiC Particle Cluster Strength

Yang et al. [33] used Rump's model to calculate the tensile strength of a SiC agglomerate using Equation (5). In Equation (5), T_{max} is the max tensile strength of the SiC cluster, φ is the volume fraction of SiC particles in the cluster, a is the average radius of the SiC particles, F is the average binding force of a single bond, and n_b is the average number of bonds per particle.

$$T_{max} = \frac{9}{32}\frac{\varphi}{\pi a^2} n_b F \tag{5}$$

The particle–particle cohesive force was in the range of 5 nN to 15 nN for different ceramic particles of similar sizes, as reported by Jones et al. [34]. The work by Jones et al. has already been used as a reference by Yang et al. [19] when investigating the influence of reinforcing particle distributions on the casting characteristics of Al-SiCp composites. ϕ is the maximum packing fraction calculated by Equations (6) and (7), where r_p is the aspect ratio of the SiC particles—these equations approximate the Kitano [35] and Pabst [36] models.

$$\Phi_{SiC} = \frac{2}{0.321 \times r_p + 3.02} \tag{6}$$

$$\frac{n_{SiC}}{n_{fcc}} = \frac{\phi_{SiC}}{\phi_{fcc}} \tag{7}$$

In Equation (7), n_{fcc} is the maximum number of bonds for an fcc structure, which is 12; n_{SiC} is the maximum number of bonds in a SiC cluster; ϕ_{SiC} is the maximum SiC particle package density, and ϕ_{fcc} is the packing density of an fcc structure, which is 0.74. After the calculation of Equations (6) and (7), the maximum number of bonds per particle, n_{SiC}, is shown in Table 4, together with the max tensile strength of the SiC particle cluster and the data used in the calculations. The larger particle sizes exhibit a smaller tensile strength, which is reasonably similar to the work by Tzantzis et al. [37].

Table 4. The calculated maximum strength of particle clusters.

MMC	C0_23	C0_50	C0_10	C0_mix	C1_23
r_p	1.86 ± 0.023	1.81 ± 0.044	1.75 ± 0.036	1.72 ± 0.011	1.73 ± 0.014
ϕ_{SiC}	0.5529	0.5553	0.5583	0.5598	0.5593
n_{SiC}	8.97	9.01	9.05	9.08	9.07
a [μm]	7 ± 0.45	16 ± 0.70	6 ± 0.90	9.5 ± 0.75	7 ± 0.55
T_{max} [Pa]	44.45–136.37	8.7–26.22	62.47–187.43	24.98–74.96	45.98–137.98

3.2.2. The Shear Stress Generated by the Stirring Device

The rotor–stator stirring device could generate intensive shearing, and the shear force should break up the particle clusters. While the shear force in the rotor–stator device could not be directly measured, the τ_{MMC} could be calculated from Equations (2)–(4).

Figure 5 shows the relationship between torque and rotation speed in the rotor–stator device. The torque varied as the rotation speed increased.

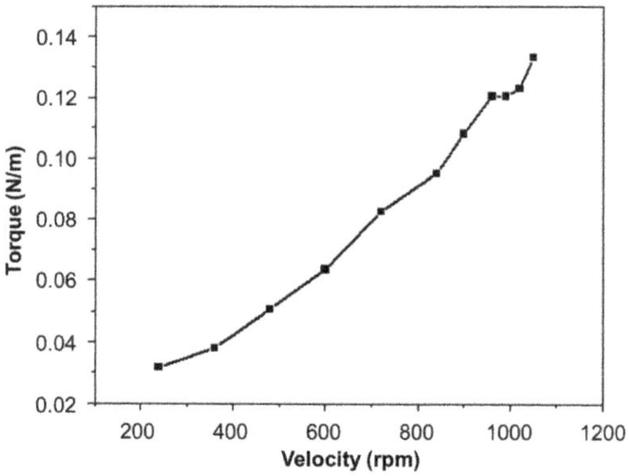

Figure 5. The relationship between the torque and the rotation speed in the rotor–stator device used for water modeling.

According to Figure 5, the stirring velocity of 500 rpm corresponded to a torque, T, calculated as 0.051 N×m. Using the viscosity, μ_{MMC} [Pa×s], of the Al-Si/SiCp slurry and the velocity gradient, $\partial u/\partial z$, the shear strength, τ_{MMC}, of the stator during the stirring of the composite slurry was calculated using Equation (4). According to Guth and Simha [29], reported by Yang et al. [33], the viscosity of the Al-Si/SiCp slurry can be calculated from Equation (8) [29], where Φ is the volume fraction of SiC particles shown in Table 2 and μ_0 = 1.4 mPa × s is the dynamic viscosity of the Al-10Si-Mg melt [38]:

$$\mu = \mu_0 \left(1 + \frac{5}{2}\Phi + \frac{109}{14}\Phi^2\right). \quad (8)$$

The shear stress in the holes through the stator wall was calculated from Equation (9) [19]:

$$\tau_H = \frac{f\rho\vartheta^2}{2}. \quad (9)$$

The f is the Fanning friction factor, which is 0.02 for a turbulent flow in a smooth steel pipe [33], ρ is the flow density, and θ is the mean flow velocity. The Flow-3D simulations approximated the flow velocity, θ, to be 8.15 m/s. The composite density was calculated using the rule of mixtures, shown in Equation (10).

$$\rho_c = V_p\rho_p + V_m\rho_m \quad (10)$$

In Equation (10), ρ_c is the density of the composite material, V_p and V_m are the volume fractions of the SiC particles and matrix alloy, and ρ_p and ρ_m are the density values of the SiC particles and the matrix alloy, respectively. The density of the matrix alloy used is 2.7 g/cm^3 [39], and the density of the SiC particles is 3.2 g/cm^3 [40]. The parameters and calculated data are shown in Table 5. It should be noted that the given speed of 8.15 m/s is an estimate based on the simulations and will be slightly lower in reality. However, the viscosity variations in the added particle ranges in this study will produce a viscosity variation less than the uncertainty of the viscosity variations of the aluminum melt itself, as described by Dinsdale and Quested [41].

Table 5. The parameters that were used to calculate the shear strength and the calculated result.

MMC	C0_23	C0_50	C0_10	C0_Mix	C1_23
μ [Pa·s]	2.10×10^{-3}	2.46×10^{-3}	1.56×10^{-3}	2.00×10^{-3}	2.10×10^{-3}
τ_{MMC} [kPa]	4.02	4.91	3.58	3.99	4.50
ρ [kg/m^3]	2743	2776	2724	2743	2762
τ_H [kPa]	1.80	1.82	1.78	1.80	1.81

In conclusion, the shear stress of the rotor–stator device and the holes through the stator wall was 1.8–5.5 kPa. This shear stress exceeds the shear strength of the SiC clusters, and it may thus be assumed that the shear stress in the rotor–stator region was sufficient to break up the SiCp clusters. This finding is further supported by the fact that Yang et al. [33] calculated the tensile strength of a SiC cluster to be in the range of 6.5–19.5 Pa. It is thus reasonable to conclude that the rotor–stator configuration in the current study would have the ability to fracture a SiC cluster to facilitate the dispersion of individual particles. Kahl et al. [42] evaluated the tensile strength of the 0.2 μm-thick oxide skin to be approximately $\sigma_{max} \approx 5$ MPa. This value is lower than other researchers found, with results ranging from 20 MPa to 250–300 MPa [43]. Due to the high strength of the oxide films, the shear stress generated by the rotor–stator device in the current study would not rupture the oxide skin. If the oxide skin is wrapped around the clusters as they break the top surface during the additions, that would present a problem as the rotor–stator system would be unable to break the cluster oxide skin. This would prevent the air inside the clusters from leaving and aid in the floatation of the SiCp clusters, reducing the transfer efficiency.

3.3. The Flow Pattern in the Furnace

Figure 6 shows the simulation result of the dynamic flow pattern of the slurry in the furnace. The black dots represent the SiC particles based on a Lagrangian formulation for the tracer particles.

The open bottom of the stirring head acts as a pump as it sucks in the slurry and pushes it out through the stator wall holes. Two main circulation regions form patterns similar to pseudo-caverns [44]. The stirring head drives these two flows, with one flow directed upwards, forming pseudo-caverns (A), and another downwards, forming pseudo-caverns (B). The SiC particles follow these two pseudo-caverns moving in the furnace. The different colors label the slurry velocity. The low-velocity zone located at the top of the furnace keeps the fluid level steady and does not allow gas entrance into the melts. In the high-velocity zone inside the stator, the high-velocity gradient will generate an intensive shear strength, so the high shear strength is located at that site. A SiC cluster usually floats up and stays in the low-velocity zone, so putting an SiC cluster into the stirring head will affect the transfer efficiency and particle dispersion.

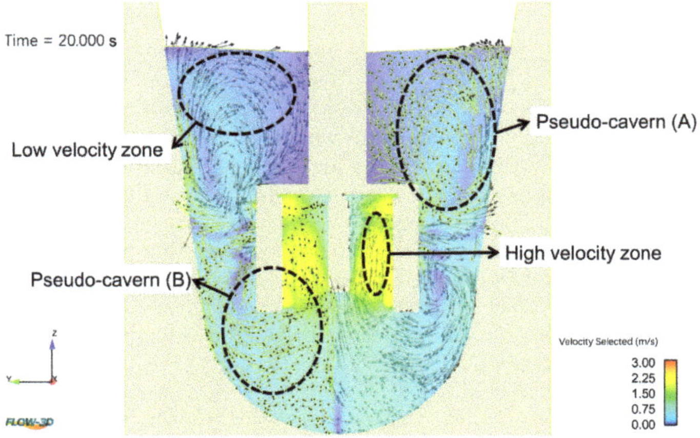

Figure 6. The rotor–stator device stirring Flow-3D simulation (500 rpm).

3.4. Particle Dispersion Analysis

Starting with a qualitative description of the dispersion, the following is observed. Figure 7 shows the morphology of the SiC particles in the composite materials. Figure 7a,e show the 23 μm-particle-reinforced composite, and the SiC particles are homogeneously dispersed in these two composites. Figure 7b shows the 50 μm-particle-reinforced composite, the largest particle size in the matrix. There is no apparent agglomeration observed. Figure 7c shows the 10 μm-particle-reinforced composite, the smallest particle size used in this study, and the SiC particles are easily found in clusters. In Figure 7c, there are traces of oxide films, indicated with an arrow, supporting the hypothesis that clusters may be promoted by oxides, hindering the wetting of the clusters. The idea is that the air inside the cluster is trapped by the oxide film wrapped around the SiC particles and makes the cluster float to the top regions of the melt, where the flow is quiescent. This would then hinder the disintegration of the floated clusters, reducing the transfer efficiency. Figure 7d shows the mixed-particle-reinforced composite, where 10 μm-size particles form satellite particles around the larger sized 23 μm and 50 μm particles and promote the formation of SiC clusters.

The equivalent diameter and average neighbor distance of the clusters were measured, among other geometrical features. Figure 8 represents the image analysis result considering only the clustered SiC particles, neglecting the well-dispersed ones appearing in the matrix alloy.

Table 6 shows the measured values forSiC particle clusters. The average neighbor distance in the materials C0_23 and C1_23 is 1.46 ± 0.81 mm and 1.18 ± 0.86 mm, respectively. All the investigated clusters had an area larger than 0.01 mm^2. The material C0_10 shows the smallest average neighbor distance of 0.99 ± 0.71 mm, highlighting the lower quality dispersion of the SiC particles compared to the other composites. The material C0_mix showed a 1.13 ± 0.84 mm average neighbor distance. Due to the addition of 10 μm-sized particles, the dispersion decreased.

The use of large-sized particles reduced the clustering of SiC particles by almost 80%, according to Table 6. The materials C0_23 and C1_23 contain identical particles, and the probability of clusters is expected to be similar for both materials. The fraction of clustered

SiC particles reported in Table 6 for the two materials ranges from 10% to 20%. The different matrix alloy resulted in a 53.7% increase in clustered SiC particles, from 13.7% to 21.1%.

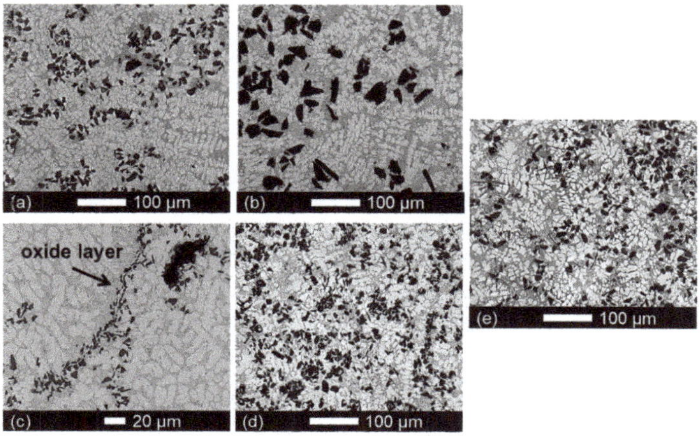

Figure 7. The morphology of SiC particles in the composite: (**a**) material C0_23; (**b**) material C0_50; (**c**) material C0_10, where the arrow indicates the residues of an oxide film; (**d**) material C0_mix; (**e**) material C1_23.

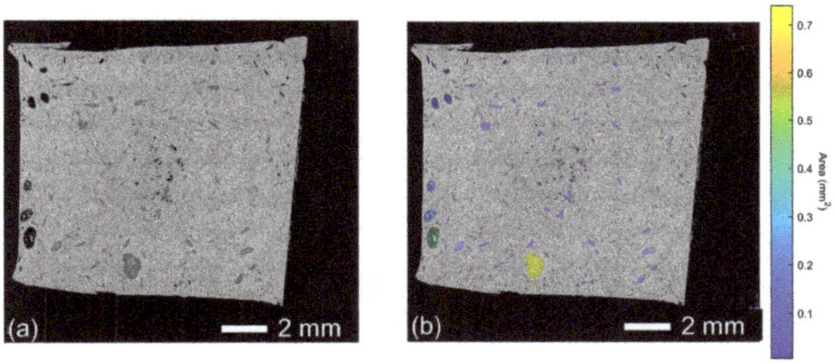

Figure 8. The method used to analyze the SiC clusters (material C1_23): (**a**) before and (**b**) after cluster detection in the MIPAR software.

Table 6. The results of the image analysis on SiCp clusters.

MMC	C0_23	C0_50	C0_10	C0_mix	C1_23
Average neighbor distance [mm]	1.46 ± 0.81	3.95 ± 2.56	0.99 ± 0.71	1.13 ± 0.84	1.18 ± 0.86
Area fraction of SiC particles	7.5%	13.1%	4.1%	7.2%	10.7%
Area fraction of SiCp clusters	1.9%	0.7%	2.8%	2.5%	4.0%
Fraction of clustered SiC particles	13.7%	2.9%	37.7%	19.0%	21.1%
Variation in clustered SiC particles compared to C0_23		−78.8%	+174.4%	+38.6%	+53.7%

The degree of clustering is a critical factor, starting with the clustering tendencies after the thermal pre-treatment and continuing with the agglomerating tendency as they pass through the top surface of the molten metal. The possibility of being coated by an aluminum oxide film when added to the melt can significantly affect the transfer efficiency. These observations are summarized in Figure 9, where the transfer efficiency is plotted against the degree of clustering.

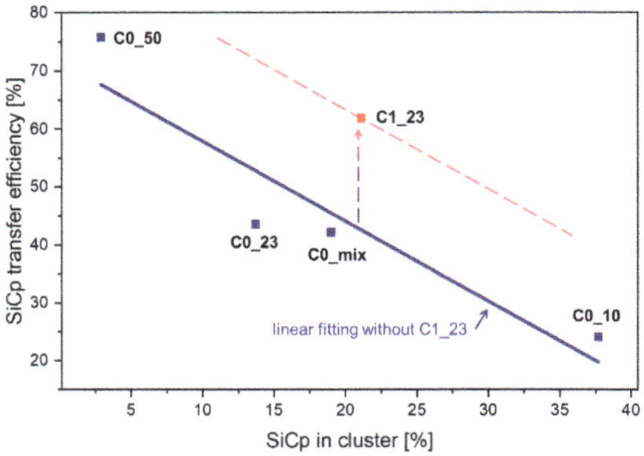

Figure 9. The transfer efficiency versus the degree of clustering confirms that the clustering level and the chemical composition of the matrix alloy influence the transfer efficiency. The linear fitting (blue solid line) on the C0 materials and its rigid translation (red dashed line) to the C1 material are meant to highlight the role of the different matrix alloys.

Despite the low value of the regression coefficient (0.87), the linear fitting helps visualize the role of the matrix alloy in the clustering and transferring of SiC particles. A low transfer efficiency is directly related to a high clustering tendency and a reinforcement loss. The red dashed line (refer to colored version) is a rigid translation of the blue fitting line and is added to visualize the possible effect of Ce and La: adding rare earth elements to the matrix alloy increased the transfer efficiency but decreased the particle dispersion. The increased wettability increases the transfer efficiency, but the explanation of the decreased particle dispersion is still somewhat unclear as the reaction has possible effects on both the melt and oxides surrounding the clusters.

4. Conclusions

The current study focuses on an Al-MMC solution for a challenging lightweight application such as brake discs for high-speed trains. The required high-temperature performance for such an application may be addressed by an Al-MMC with rare earth element additions such as Ce and La to improve the strength of the matrix alloy [22]. The addition of SiC as a reinforcing material and rare earth elements affects the material and component fabrication. This paper aimed to investigate what hinders a high transfer efficiency of SiC particles to an Al melt and a good dispersion with a low level of clustering to allow efficient and cost-effective manufacturing. The following observations were made:

- The required heat treatment to support wetting results in a certain degree of clustering in the SiC particlesafter the thermal pre-treatment. The presence of smaller particles aggravated the degree of clustering.
- The transfer efficiency increased with increasing particle sizes. The mechanism suggested for this relationship was that the clusters from the pre-treatment were encapsulated by an oxide film during addition, making them buoyant. The pumping action of the rotor–stator device left the top surface quiescent, making it difficult to reintroduce floated clusters and thus reducing the transfer efficiency. The SiCp clusters should break up once entering the stirring head, but the shear stress in the stirring heads was insufficient to break the aluminum oxide skins generated during the melting process.
- Adding Ce and La increased the transfer efficiency. La, having a high oxygen affinity, could potentially affect the wetting reaction between SiO_2, which has a stronger driving force to react with oxygen than Mg. However, this effect on wetting would need further investigation to be fully understood and described.

Author Contributions: Conceptualization, A.E.W.J. and J.Z.; methodology, A.D., L.L. and A.E.W.J.; investigation, A.D. and L.L.; resources, A.E.W.J. and J.Z.; writing—original draft preparation, A.D. and L.L.; writing—review and editing, all authors together; supervision, A.E.W.J., J.Z., G.Y. and K.W.; project administration, A.E.W.J. and J.Z.; funding acquisition, A.E.W.J., J.Z. and L.L. All authors have read and agreed to the published version of the manuscript.

Funding: This research was funded by the Short-term Recruitment Program of Foreign Expert (China), grant 20181215, and the Stiftelsen för Kunskaps-och Kompetensutveckling (Sweden), Prospekt grant ProForAl-20201702.

Data Availability Statement: The data presented in this study are available on request from the corresponding author. The data are not publicly available due to ongoing research and proprietary information.

Acknowledgments: The authors gratefully acknowledge the technical staff at the Department of Materials and Manufacturing of the School of Engineering, Jönköping University. Ansys is gratefully acknowledged for its support with Granta Selector 2023 R1 software through the Research Partnership with Tekniska Högskolan i Jönköping.

Conflicts of Interest: The authors declare no conflict of interest. The funders had no role in the design of the study; in the collection, analyses, or interpretation of data; in the writing of the manuscript; or in the decision to publish the results.

References

1. Huczko, A.; Dabrowska, A.; Savchyn, V.; Popov, A.I.; Karbovnyk, I. Silicon Carbide Nanowires: Synthesis and Cathodoluminescence. *Phys. Status Solidi (b)* **2009**, *246*, 2806–2808. [CrossRef]
2. Xu, H.; Li, X.; Tong, Z.; Zhang, B.; Ji, H.; Cannilla, C.; Vakros, J.; Hapeshi, E.; Xu, H.; Li, X.; et al. Thermal Radiation Shielding and Mechanical Strengthening of Mullite Fiber/SiC Nanowire Aerogels Using In Situ Synthesized SiC Nanowires. *Materials* **2022**, *15*, 3522. [CrossRef] [PubMed]
3. Lebedev, A.S.; Suzdal'tsev, A.V.; Anfilogov, V.N.; Farlenkov, A.S.; Porotnikova, N.M.; Vovkotrub, E.G.; Akashev, L.A. Carbothermal Synthesis, Properties, and Structure of Ultrafine SiC Fibers. *Inorg. Mater.* **2020**, *56*, 20–27. [CrossRef]
4. Zheng, X.; Lee, H.; Weisgraber, T.H.; Shusteff, M.; DeOtte, J.; Duoss, E.B.; Kuntz, J.D.; Biener, M.M.; Ge, Q.; Jackson, J.A.; et al. Ultralight, Ultrastiff Mechanical Metamaterials. *Science* **2014**, *344*, 1373–1377. [CrossRef] [PubMed]
5. Thomas, A.; Zervos, N.; Ekelund, A.; Awe, S.A. Simulation Study on the Thermomechanical Behaviour of Al-MMC Automotive Brake Discs. In Proceedings of the Eurobrake 2019, Dresden, Germany, 21–23 May 2019; pp. 1–12.
6. European Vehicle Emissions Standards—Euro 7 for Cars, Vans, Lorries and Buses. Available online: https://ec.europa.eu/info/law/better-regulation/have-your-say/initiatives/12313-European-vehicle-emissions-standards-Euro-7-for-cars-vans-lorries-and-buses_en (accessed on 23 June 2022).
7. Serrenho, A.C.; Norman, J.B.; Allwood, J.M. The Impact of Reducing Car Weight on Global Emissions: The Future Fleet in Great Britain. *Philos. Trans. R. Soc. A Math. Phys. Eng. Sci.* **2017**, *375*, 20160364. [CrossRef] [PubMed]
8. Kenworthy, J.R. Transport Energy Use and Greenhouse Gases in Urban Passenger Transport Systems: A Study of 84 Global Cities. In Proceedings of the International Sustainability Conference 2003, Fremantle, Western Australia, 17–19 September 2003; pp. 1–28.
9. Hredzak, B.; Gair, S.; Eastham, J.F. Control of an EV Drive with Reduced Unsprung Mass. *IEE Proc. Electr. Power Appl.* **1998**, *145*, 600. [CrossRef]
10. Lattanzi, L.; Etienne, A.; Li, Z.; Manjunath, T.; Nixon, N.; Jarfors, A.E.W.; Awe, S.A. The Influence of Ni and Zr Additions on the Hot Compression Properties of Al-SiCp Composites. *J. Alloys Compd.* **2022**, *905*, 164160. [CrossRef]
11. Skibo, M.D.; Schuster, D.M. Process for Preparation of Composite Materials Containing Nonmetallic Particles in a Metallic Matrix. U.S. Patent 4,865,806, 12 September 1989.
12. Bao, S.; Tang, K.; Kvithyld, A.; Engh, T.; Tangstad, M. Wetting of Pure Aluminium on Graphite, SiC and Al_2O_3 in Aluminium Filtration. *Trans. Nonferrous Met. Soc. China* **2012**, *22*, 1930–1938. [CrossRef]
13. Lloyd, D.J. Particle Reinforced Aluminium and Magnesium Matrix Composites. *Int. Mater. Rev.* **1994**, *39*, 1–23. [CrossRef]
14. Hashim, J.; Looney, L.; Hashmi, M.S.J. Metal Matrix Composites: Production by the Stir Casting Method. *J. Mater. Process. Technol.* **1999**, *92–93*, 1–7. [CrossRef]
15. Malaki, M.; Tehrani, A.F.; Niroumand, B.; Gupta, M. Wettability in Metal Matrix Composites. *Metals* **2021**, *11*, 1034. [CrossRef]
16. Ureña, A.; Martínez, E.E.; Rodrigo, P.; Gil, L. Oxidation Treatments for SiC Particles Used as Reinforcement in Aluminium Matrix Composites. *Compos. Sci. Technol.* **2004**, *64*, 1843–1854. [CrossRef]
17. Shanmuga Priyan, V.G.; Kanmani, S. Effect of Ultrasonic Treatment during Stir Casting on Mechanical Properties of AA6063-SiC Composites. *Mater. Chem. Phys.* **2023**, *294*, 126977. [CrossRef]
18. Fan, Z.Y.; Zuo, Y.B.; Jiang, B. A New Technology for Treating Liquid Metals with Intensive Melt Shearing. *Mater. Sci. Forum* **2011**, *690*, 141–144. [CrossRef]

19. Yang, X.; Barekar, N.S.; Ji, S.; Dhindaw, B.K.; Fan, Z. Influence of Reinforcing Particle Distribution on the Casting Characteristics of Al-SiCp Composites. *J. Mater. Process. Technol.* **2020**, *279*, 116580. [CrossRef]
20. Capes, C.E. The Correlation of Agglomerate Strength with Size. *Powder Technol.* **1972**, *5*, 119–125. [CrossRef]
21. Wilks, G.B. The Influence of Reinforcement Homogeneity on the Deformation and Fracture of Discontinuously Reinforced Aluminum Matrix Composite. Ph.D. Thesis, Pennsylvania State University, Graduate School College of Earth and Mineral Sciences, University Park, PA, USA, 2007.
22. Du, A.; Jarfors, A.E.W.; Zheng, J.; Wang, K.; Yu, G. The Influence of La and Ce on Microstructure and Mechanical Properties of an Al-Si-Cu-Mg-Fe Alloy at High Temperature. *Metals* **2021**, *11*, 384. [CrossRef]
23. Du, A.; Lattanzi, L.; Jarfors, A.W.E.; Zheng, J.; Wang, K.; Yu, G. On the Hardness and Elastic Modulus of Phases in Sic-Reinforced al Composite: Role of La and Ce Addition. *Materials* **2021**, *14*, 6287. [CrossRef]
24. Du, A.; Lattanzi, L.; Jarfors, A.E.W.; Zhou, J.; Zheng, J.; Wang, K.; Yu, G. The Influence of Ce, La, and SiC Particles Addition on the Formability of an Al-Si-Cu-Mg-Fe SiCp-MMC. *Materials* **2022**, *15*, 3789. [CrossRef]
25. Jarfors, A.E.W.; Ghasemi, R.; Awe, S.; Jammula, C.K. Comparison between High-Pressure Die-Cast and Rheo-Cast Aluminium SiCp MMC; Wear and Friction Behaviour. *La Metall. Ital.* **2021**, *12*, 13–18.
26. Awe, S.A.; Thomas, A. The Prospects of Lightweight SICAlight Discs in the Emerging Disc Brake Requirements. In Proceedings of the Eurobrake 2021, Online, 18–20 May 2021; pp. 1–6.
27. Du, A.; Lattanzi, L.; Jarfors, A.E.W.; Zheng, J.; Wang, K.; Yu, G. Role of Matrix Alloy, Reinforcement Size and Fraction in the Sliding Wear Behaviour of Al-SiCp MMCs against Brake Pad Material. *Wear* **2023**, *530*, 204969. [CrossRef]
28. Bird, B.R.; Stewart, W.E.; Lightfoot, E.N. *Transport Phenomena*, 2nd ed.; John Wiley & Sons: Hoboken, NJ, USA, 2006; ISBN 978-0-470-11539-8.
29. Guth, E.; Simha, R. Untersuchungen Über Die Viskosität von Suspensionen Und Lösungen. 3. Über Die Viskosität von Kugelsuspensionen. *Kolloid-Zeitschrift* **1936**, *74*, 266–275. [CrossRef]
30. Campbell, J. The Melt. In *Castings*; Elsevier: Amsterdam, The Netherlands, 2003; pp. 1–16.
31. Liu, Q.; Wang, F.; Qiu, X.; An, D.; He, Z.; Zhang, Q.; Xie, Z. Effects of La and Ce on Microstructure and Properties of SiC/Al Composites. *Ceram. Int.* **2020**, *46*, 1232–1235. [CrossRef]
32. Hasegawa, M. Ellingham Diagram. *Treatise Process Metall.* **2014**, *1*, 507–516. [CrossRef]
33. Yang, X.; Huang, Y.; Barekar, N.S.; Das, S.; Stone, I.C.; Fan, Z. High Shear Dispersion Technology Prior to Twin Roll Casting for High Performance Magnesium/SiCp Metal Matrix Composite Strip Fabrication. *Compos. Part A Appl. Sci. Manuf.* **2016**, *90*, 349–358. [CrossRef]
34. Jones, R.; Pollock, H.M.; Geldart, D.; Verlinden, A. Inter-Particle Forces in Cohesive Powders Studied by AFM: Effects of Relative Humidity, Particle Size and Wall Adhesion. *Powder Technol.* **2003**, *132*, 196–210. [CrossRef]
35. Kitano, T.; Kataoka, T.; Shirota, T. An Empirical Equation of the Relative Viscosity of Polymer Melts Filled with Various Inorganic Fillers. *Rheol. Acta* **1981**, *20*, 207–209. [CrossRef]
36. Pabst, W.; Gregorová, E.; Berthold, C. Particle Shape and Suspension Rheology of Short-Fiber Systems. *J. Eur. Ceram. Soc.* **2006**, *26*, 149–160. [CrossRef]
37. Tzamtzis, S.; Barekar, N.S.; Hari Babu, N.; Patel, J.; Dhindaw, B.K.; Fan, Z. Processing of Advanced Al/SiC Particulate Metal Matrix Composites under Intensive Shearing—A Novel Rheo-Process. *Compos. Part A Appl. Sci. Manuf.* **2009**, *40*, 144–151. [CrossRef]
38. Wei, P.; Wei, Z.; Chen, Z.; He, Y.; Du, J. Thermal Behavior in Single Track during Selective Laser Melting of AlSi10Mg Powder. *Appl. Phys. A* **2017**, *123*, 604. [CrossRef]
39. Al-10Si alloy, Granta Selector 2023 R1, Version 23.1.1, consulted in August 2023.
40. Silicon carbide particles (SiCp), Granta Selector 2023 R1, Version 23.1.1, consulted in August 2023.
41. Dinsdale, A.T.; Quested, P.N. The Viscosity of Aluminium and Its Alloys—A Review of Data and Models. *J. Mater. Sci.* **2004**, *39*, 7221–7228. [CrossRef]
42. Kahl, W.; Fromm, E. Examination of the Strength of Oxide Skins on Aluminum Alloy Melts. *Metall. Trans. B* **1985**, *16*, 47–51. [CrossRef]
43. Engler, S.; Ellerbrok, R. Über Das Formfüllungsvermögen Der Gusswerkstoffe. *Giessereiforschung* **1974**, *26*, 49–62.
44. Doucet, L.; Ascanio, G.; Tanguy, P.A. Hydrodynamics Characterization of Rotor-Stator Mixer with Viscous Fluids. *Chem. Eng. Res. Des.* **2005**, *83*, 1186–1195. [CrossRef]

crystals

MDPI

Article

Combined Effect of Particle Reinforcement and T6 Heat Treatment on the Compressive Deformation Behavior of an A357 Aluminum Alloy at Room Temperature and at 350 °C

Sarah Johanna Hirsch *, Nadja Berndt, Thomas Grund and Thomas Lampke

Institute of Materials Science and Engineering, Chemnitz University of Technology, Erfenschlager Str. 73, 09125 Chemnitz, Germany; nadja.berndt@mb.tu-chemnitz.de (N.B.); thomas.grund@mb.tu-chemnitz.de (T.G.); thomas.lampke@mb.tu-chemnitz.de (T.L.)

* Correspondence: sarah-johanna.hirsch@mb.tu-chemnitz.de

Abstract: Solid state sintering of cast aluminum powders by resistance heating sintering (RHS), also known as spark plasma sintering or field-assisted sintering technique, creates a very fine microstructure in the bulk material. This leads to high performance material properties with an improved strength and ductility compared to conventional production routes of the same alloys. In this study, the mechanical behavior of an RHS-sintered age-hardenable A357 (AlSi7Mg0.6) cast alloy and a SiC_p/A357 aluminum matrix composite (AMC) was investigated. Aiming for high strength and good wear behavior in tribological applications, the AMC was reinforced with a high particle content (35 vol.%) of a coarse particle fraction (d_{50} = 21 µm). Afterwards, separated and combined effects of particle reinforcement and heat treatment were studied under compressive load both at room temperature and at 350 °C. At room temperature compression, the strengthening effect of precipitation hardening was about twice as high as that for the particle reinforcement, despite the high particle content. At elevated temperatures, the compressive deformation behavior was characterized by simultaneously occurring temperature-activated recovery, recrystallisation and precipitation processes. The occurrence and interaction of these processes was significantly affected by the initial material condition. Moreover, a rearrangement of the SiC reinforcement particles was detected after hot deformation. This rearrangement lead to a homogenized dispersion of the reinforcement phase without considerable particle fragmentation, which offers the potential for secondary thermo-mechanical processing of highly reinforced AMCs.

Keywords: aluminum cast alloy; AMC; compression; field-assisted sintering; heat treatment; hot deformation; mechanical properties; microstructure; SiC; solid-state sintering

1. Introduction

Al-Si-Mg cast alloys like A356 (AlSi7Mg0.3) and A357 (AlSi7Mg0.6) are used in many commercial engineering applications such as pistons, engine blocks, brake saddles, pumps, impellers and valve components [1]. Key factors for the use of these alloys are their high specific strength [1,2] and high toughness [2], particularly in their heat-treated conditions [3]. In addition, these alloys have an excellent castability [1], and due to the good fluidity in the molten state [4] they are popular candidates for additive manufacturing processes [4–6]. Nevertheless, the mechanical properties are significantly influenced by the manufacturing process, the resulting microstructure and the subsequent heat treatment. For example, the use of powder metallurgical processes can avoid potential problems of conventional cast structures, such as columnar grain growth and heterogeneous microstructures and thus gradual changes in the properties [2] due to thermal gradients and different local solidification velocities. In addition, the presence of the alloying elements Si and Mg allows for a significant improvement in strength due to precipitation hardening. The precipitation sequence is generally accepted as [7–9]: super saturated solid solution → clusters → Guinier–Preston zones (GP I, coherent) → β'' (semi-coherent, needle-shaped) → β' (semi-coherent, rod-shaped) → β (incoherent). The highest strength can be achieved by a T6 heat treatment routine, i.e., solution annealing, water quenching and artificial ageing at elevated temperatures to achieve a high amount of finely dispersed β'' precipitates.

By reinforcing an aluminum alloy with a high ceramic particle content ($\geq$20 vol.%, [10]) and particle sizes in the micrometer range, an aluminum matrix composite (AMC) with significantly improved wear behavior compared to the unreinforced monolithic aluminum matrix is created. There are many studies regarding the wear behavior of AMCs [11–16]. Beside this, the stiffness, strength, thermal stability and creep resistance are improved [10,17,18] by adding a ceramic reinforcement phase. Since up to about 60% of weight can be saved compared to grey cast iron, this material is predestined for tribologically stressed components, such as in brake discs [10,19]. Compared to wrought aluminum alloys hypereutectic and hypoeutectic (like AlSi7Mg alloys) are the predominantly used brake disc matrix materials [19], and (particulate) SiC (abbreviated: SiC_p) is the most commonly used reinforcement phase [19].

In recent decades, the mechanical behavior of AMCs under compression has been extensively studied. Mondal et al. [20] and Ye et al. [21] investigated the compression deformation behavior of Al/SiC composites with different contents of SiC_p [20] and particle sizes [21] at different strain rates. Sajjadi et al. [22] found that the compression strength is increased with increasing reinforcement content. The deformation behavior at elevated temperatures was also investigated in many studies [23–28], since required AMC properties and microstructures are often generated by the thermo-mechanical treatment of semi-finished AMC products.

It is well known that the work hardening effect during compression is, in general, significantly reduced at higher deformation temperatures. This results from the thermal activation of micromechanical processes, e.g., dislocation climb and cross slip. In addition, dynamic recovery (DRV) and dynamic recrystallisation (DRX) lead to an enhanced softening effect. DRV can be described as follows: with increased deformation temperature and low strain rate, the dislocation density decreases and the sub-grain boundaries are stronger developed [24]. Therefore, large angle grain boundaries are formed and the sub-grain size gradually increases. These tendencies can be attributed to the resulting higher thermal activation energy and the sufficient time required (for low strain rates) to rearrange the dislocations and ensure sufficient migration of the atoms, which results in flow softening [24]. According to Khodabakhshi et al. [29], who investigated Al matrix hybrid nanocomposites, the relatively high stacking fault energy of Al and its alloys in the low temperature range should facilitate the development of a subgrain structure within (high-angle) grain, which is typical for DRV. In addition to DRV, a microstructure of fine recrystallized grains with a high proportion of high-angle grain boundaries can also be formed during the deformation at higher temperatures [29]. In the progress of this DRX, new grains are permanently formed to compensate for the new, continuously generated dislocations. This dynamic mechanism is commonly observed in hot working at temperatures above 50% of the melting temperature [30]. Concerning the discussion about the materials' behavior at elevated temperatures, dynamic precipitation (DP) also needs to be considered [29,31]. DP follows a similar sequence as static precipitation and also depends on the temperature [32]. It is a diffusion-driven process supported by vacancies and dislocations [33]. The dislocations are favored nucleation sites for precipitates. As a result of dislocation movement during deformation, a homogeneous and dense distribution of very fine precipitates, which contribute to the high strength and hardness values, can be observed [34].

The deformation behavior of AMCs has been studied for a wide range of compositions. Varied material parameters were both matrix material and reinforcement phase (in terms of size, chemical composition and volume content). Though studies were conducted for a wide range of temperatures and strain rates, they usually considered AMCs with a reinforcement grade of typically below 20 vol% [23–25,28] or with wrought aluminum matrix alloys [26,27]. In the present study, the mechanical properties and microstructural evolution are investigated for a SiC_p/A357 AMC with high reinforcement grade and coarse fraction (35 vol.%, d_{50} = 21 µm), suited for high-wear applications. Mechanical tests are conducted under compressive load at room temperature and at 350 °C and compared to the unreinforced A357. In addition, the (combined) effect of two types of reinforcement phase is investigated. Firstly, it is the dispersed SiC_p phase, and secondly a static as well as dynamic precipitation reinforcement. Special focus is placed on DRX behavior and particle fracture during deformation, which are important factors that determine the mechanical properties of the AMC after a possible secondary thermo-mechanical processing.

2. Materials and Methods

2.1. Materials and Processing

An AlSi7Mg0.6 (EN AC 42.200, see DIN EN 1706 [35]) equal to A357 cast aluminum (see The Aluminum Association) powder from ECKA Granules GmbH (Velden, Germany) with an average powder size d_{50} of 40 µm was used as matrix powder. This gas-atomized powder (spherical shape) was mechanically mixed with 35 vol.% polygonal SiC_p (ESK-SIC GmbH, Ferchen, Germany) in a three-dimensional tumbler mixer (TURBULA; Willy A. Bachofen GmbH, Nidderau-Heldenbergen, Germany) for 10 min. The SiC_p fraction (size: F320) was sieved with a mesh size of 20 µm. A final SiC powder size of d_{50} = 21 µm and a distributional width of 21 µm was achieved. The AMC mixed powder fraction and the as-received A357 powder fraction were consolidated in the solid state to cylindrically shaped samples (D = 40 mm, h = 8 mm) by RHS with a sintering furnace KCE FCT-HP D 25-SI (FCT Systeme GmbH, Frankenblick, Germany). During this sintering process, the powder was heated by its own electrical resistance or indirectly by the used graphite tool. Due to the high heating and cooling rates, sintering temperatures close to the solidification temperature as well as the high sintering pressures, the used powder can be consolidated within a very short time in the range of few minutes [36]. Based on expertise from previous investigations [2,37], the sintering process was carried out in a vacuum for 5 min (holding time) at 520 °C and at a sintering pressure of 50 MPa to achieve nearly fully dense bodies (porosity < 1%) within a very short time to avoid microstructural coarsening effects. In addition, some samples were subsequently heat-treated in order to enhance the strength. The corresponding T6 heat treatment consisted of solid solution annealing for 1.0 h at 520 °C, quenching in water and subsequent artificial aging for 2.5 h at 180 °C. Consolidation and heat treatment were carried out as described in more detail in our former study [2].

2.2. Mechanical Testing

For compression tests, cylinders with a diameter of 6 mm were cut from each sintered and heat-treated sample in direction of the sintering pressure by wire eroding. The end faces of the cylinders were machined by face turning to achieve plane surfaces and a final specimen height of 6 mm. The compression specimens were taken at a radial position halfway between the center and the outer diameter of the sintered samples.

All material conditions (as-sintered and T6 heat-treated) were investigated by compression tests at room temperature (RT) and at 350 °C at a quasi-static strain rate of 10^{-3} s^{-1} in a Zwick-Roell 100 kN universal testing device (ZwickRoell GmbH & Co. KG, Ulm, Germany) up to a compression of about 50%. Three tests were performed for each condition at the respective temperatures. To reduce friction during testing, a thermally stable copper paste was applied to the end faces of the specimens.

Inductive heating was used for compression tests at elevated temperatures. The specimen temperature was controlled by a pyrometer measuring the temperature on the lateral surface. This was painted with a non-reflecting black layer to ensure a comparable emission during temperature control. All compression specimens were preheated for about one minute and then held at 350 °C for 5 min before testing.

2.3. Microstructural Analysis

A selection of tested compression test specimens were used for microstructural analysis. The longitudinal and cross sections were prepared using conventional metallographic preparation routines. Optical micrographs were taken with an Olympus GX51 microscope (Olympus Deutschland GmbH, Evident Corporation, Tokyo, Japan) for qualitative microstructural analysis. For quantitative microstructural analysis, the software tool Materials Solution (Olympus Deutschland GmbH, Evident Corporation, Tokyo, Japan; evaluation software: Olympus Stream Motion 2.1) was used. Both sample porosity and SiC particle size distribution were monitored by digital image analysis, using grey scale thresholds that were set manually (i.e., light grey (matrix including eutectic silicon), middle grey (SiC_p) and near-black (pores). To evaluate the porosity, five micrographs were taken and for each of these, an area with a size of 250 μm × 250 μm was analyzed. Ten areas with a size of 100 μm × 100 μm obtained from individual micrographs in the centers of the cross sections were used to determine the SiC particle size distribution. The identified particle sizes correspond to the respective average particle diameters determined from the Materials Solution software. The particle size was classified in fractions with increments of 2.5 μm (see Table 1), starting at 5 μm to avoid an erroneous assignment of eutectic Si phase from the matrix, and ending at 22.5 μm, since only a few coarser particles were detected, whose number were not sufficient for an adequate statistical validation.

Table 1. Classification of the SiC_p size (average particle diameters).

Fraction	I	II	III	IV	V	VI	VII
from size (μm)	≥5.0	≥7.5	≥10.0	≥12.5	≥15.0	≥17.5	≥20.0
to size (μm)	<7.5	<10.0	<12.5	<15.0	<17.5	<20.0	<22.5

Additional electron-backscatter diffraction (EBSD) measurements were carried out on sintered and T6 heat-treated samples as well as compression specimens tested at 350 °C for the material A357. The AMC conditions could not be characterized by EBSD due to the high content of reinforcement particles (35 vol.%), since the SiC_p were considerably harder than the aluminum matrix and protruded from the polished sample surface. Due to the inclination of the sample during EBSD, this resulted in extensive shadowing effects and therefore large areas with unreliable or no information. The embedded samples were manually ground and subjected to the polishing routine given in Table 2. The final vibratory polishing step (5) was carried out for 45 min in an aqueous colloidal silica suspension consisting of 100 mL of distilled water and 50 mL of MasterMet (Buehler GmbH, Leinfelden-Echterdingen, Germany).

Table 2. Polishing routine for EBSD measurements of the A357 alloy.

Step	Polishing Cloth	Polishing Suspension	Contact Force	Duration
(1)	Struers [1], Sat	3 μm, diamond	5 N	2 × 3 min
(2)	Struers [1], Plus	1 μm, diamond	5 N	2 × 4 min
(3)	Struers [1], Chem	0.06 μm, silica	5 N	1 × 3 min
(4)	Struers [1], Floc	0.25 μm, diamond	5 N	4 × 5 min
(5)	Buehler, MasterTex	0.06 μm, silica (diluted)	-	1 × 45 min

[1] Struers GmbH, Willich, Germany.

A Neon 40 EsB field emission SEM (Carl Zeiss AG, Oberkochen, Germany) equipped with a DigiView IV EBSD camera (EDAX Inc., Mahwah, NJ, USA) was used for microstructural characterization. The EBSD patterns were collected at an acceleration voltage of 20 kV with an aperture of 120 μm. The step size for all measurements was 0.1 μm. Post-processing of the raw data in OIM Analysis (EDAX Inc., Mahwah, NJ, USA) software included a slight clean-up based on the neighbor confidence index correlation (CI < 0.05).

2.4. *Differential Scanning Calorimetry*

The precipitation behavior of selected A357 samples was analyzed using a DSC 1 System (Mettler Toledo AG, Gießen, Switzerland). Therefore, cylindrical specimens with a diameter of 5.0 mm, a height of 5.5 mm and a resulting mass of approx. 255 mg were tested. A high-purity Al5N (99.999% Al) sample was used as the reference. The samples were placed in aluminum crucibles and tested in nitrogen. Heating was carried out at a rate of 20 K/min up to 400 °C with subsequent cooling (20 K/min) to RT. This thermal cycle was iterated twice for baseline correction.

3. Results and Discussion

3.1. *Microstructural Characterization of the Sample Conditions before Compression Test*

Representative optical micrographs of the RHS-sintered SiC_p/A357 in both as-sintered and T6 heat-treated condition are presented in Figure 1a,b, respectively. In general, the microstructure of the investigated SiC_p/A357 is characterized by reinforcing particles (dark grey) surrounded by the aluminum matrix (light grey). Due to the mixed state of the fed powder as well as the short process time of RHS that does not allow strong diffusion, there are some SiC_p agglomerations that locally led to pores (black). In addition, the RHS-sintered matrix material exhibits very fine, evenly distributed eutectic Si of globular shape (see Figure 1a). This particulate phase coarsened during the T6 heat treatment (Figure 1b), as described in [2] for the unreinforced A357 alloy. It was also observed and discussed that the porosity of the sintered material increased during heat treatment [2], which can be confirmed for the AMCs by Archimedean density measurements (3D) and quantitative microstructure analysis (2D) (see Table 3). The difference to theoretical density was calculated from the determined density values. Therefore, the density was subtracted from the theoretical density (here: 2.67 $g{\cdot}cm^{-3}$ for the unreinforced condition and 2.86 $g{\cdot}cm^{-3}$ for a reinforcement content of 35 vol.% SiC).

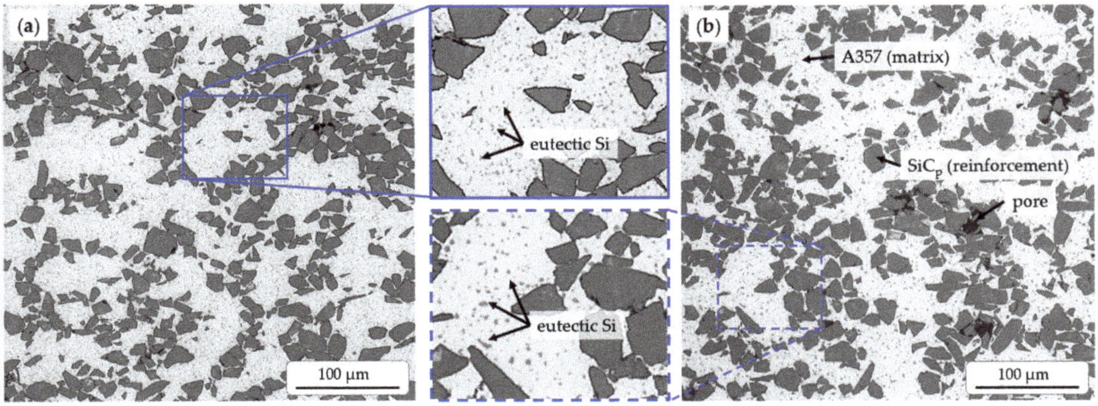

Figure 1. Optical micrographs of the RHS-sintered AMC in the as-sintered (**a**) and T6 heat-treated condition (**b**). The eutectic silicon coarsened due to T6 heat treatment (see detail views).

Table 3. The difference in theoretical density and porosity of the investigated materials.

Material	Difference to Theoretical Density (%)		Porosity (%)	
	as-Sintered	T6	as-Sintered	T6
SiC_p/A357	1.4 ± 0.0	2.1 ± 0.3	0.6 ± 0.3	1.6 ± 0.1
A357	0.1 ± 0.0 [1]	2.0 ± 0.4 [1]	0.0 ± 0.0	0.4 ± 0.1

[1] after [2].

It can be assumed that along with the trapped residual gas between former powder particles during sintering [2], argon gas that was trapped in a solid solution during rapid cooling after the gas atomization process (powder manufacturing process) may also have been released during the static (pressureless) solution annealing. Even though the porosity only provides information about the microstructure within the section plane, both the difference to the theoretical density and the porosity show that the initial porosity is higher for the AMCs. However, the increase in the difference to the theoretical density as a result of the heat treatment is stronger for the unreinforced A357 sinter material than for SiC_p/A357, which is due to the higher volume of aluminum alloy in the sintered body. The opposite effect can be observed for the porosity. In general, the quantitative microstructure analyses show smaller values for porosity, which are caused by local differences between the analyzed small areas or volumes, even if a high number of pictures are analyzed and operator influences (e.g., by differently set grey-scale thresholds) can be ruled out. On this basis, the difference to the theoretical density, which gives an average value for a larger sample area or volume, should be preferred. Accordingly, it can be concluded from these results that the materials have a comparable density regardless of the reinforcement. However, for a quantitative evaluation of very small or localized areas, only quantitative microstructure analysis can be considered, see Section 3.4.

In addition to the growth of pores and the coarsening of eutectic silicon, as reported for both investigated materials, the T6 heat treatment also promotes grain growth, as shown in Figure 2 for A357. The color-coded orientation maps were obtained from EBSD measurements on the longitudinal sections of the sintered samples and show fine globular grains for both conditions in contrast to conventionally cast AlSiMg alloys with a grain size of more than a hundred microns (see, e.g., [38]). The grain size increases during T6 heat treatment resulting in an average grain size of about 10 µm in contrast to a size of about 5 µm in the as-sintered condition. The average grain width in AlSiMg processed, e.g., via selective laser melting (or laser powder bed fusion) using powders with a similar initial powder particle size of 40 µm, is in the range of 12 µm to 14 µm [39,40]. However, the structure is often heterogeneous with small amounts of equiaxed grains and mainly columnar grains that can exhibit a strong texture [39,41] due to their directional growth during solidification. In contrast, both conditions, as sintered (Figure 2a) and T6 (Figure 2b), exhibit nearly homogeneous distributions of all crystal orientations, i.e., no texturing of the material occurs during sintering or heat treatment which would be indicated by a large number of grains with a similar color. It is to be noted that the plotted orientation colors correspond to the crystal orientations parallel to the direction of the sintering pressure, i.e., perpendicular to the observed section surface, since this would be the direction where a preferred crystal orientation could be expected first.

3.2. Compressive Behavior at RT

The results of the compressions tests are given in Figure 3. Representative stress–strain curves (Figure 3a) show a significant increase in strength due to particle reinforcement and an additional strengthening effect due to T6 heat treatment. Corresponding values for the true offset yield strength (OYS) at a plastic deformation of 0.2% and the true strength at a plastic deformation of 15% were calculated from the stress–strain curves and are given in Figure 3b for A357 and SiC_p/A357, respectively. Since a definite failure (maximum compression strength) of the specimens was not detectable, the strength at a plastic deformation of 15% was determined as an alternative comparative value. Up to this degree of deformation, there was no visible crack formation at the lateral faces of the compression specimens, and the effect of friction is comparatively low due to lubrication.

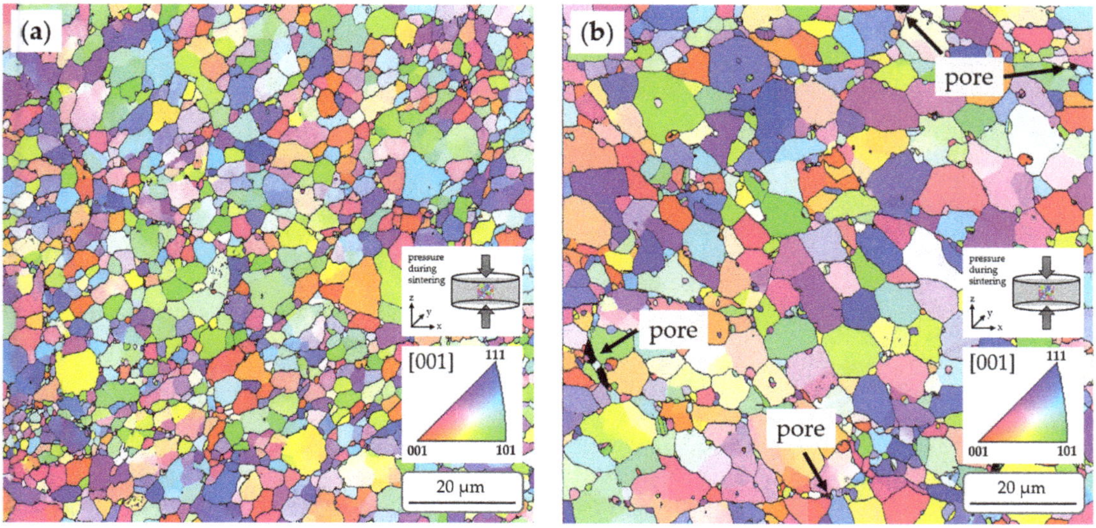

Figure 2. Color-coded orientation maps obtained by EBSD measurements on the longitudinal sections of sintered samples for (**a**) the as-sintered condition and (**b**) the T6 heat-treated condition of A357. The colors correspond to the crystal orientation parallel to the sintering pressure, i.e., parallel to the [001] sample direction. High-angle grain boundaries (>15°) are marked by black lines.

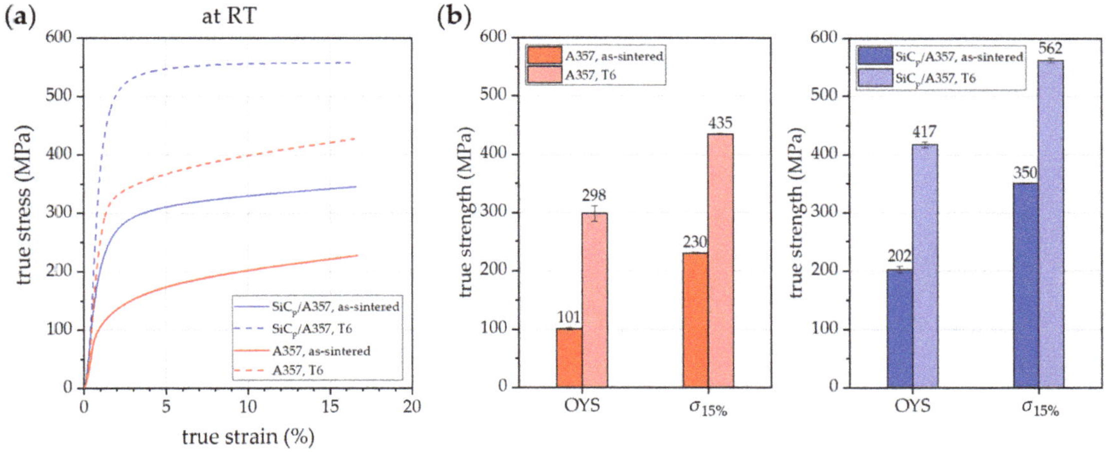

Figure 3. (**a**) Representative true stress–strain curves plotted up to a true compression of 15% and (**b**) mechanical properties obtained from compression tests at RT for the as-sintered and T6 conditions.

The as-sintered A357 shows the lowest OYS of about 100 MPa followed by the most pronounced work hardening. By reinforcing (SiC_p/A357), the OYS is doubled. The elasto-plastic compressive behavior is characterized by an initially higher rate of strain hardening compared to A357 that decreases after a plastic strain of about 3%. After T6 heat treatment, a similar work hardening behavior can be observed for both materials. However, the strength is increased significantly, i.e., the OYS is higher by about 200 MPa for both A357 and SiC_p/A357. This indicates a relative increase in yield strength of about 200% for A357 and of about 100% for SiC_p/A357. The differences regarding the OYS as well as the work hardening behavior are mainly caused by the dispersion (SiC_p) and/or precipitation hardening (T6) effects. As shown in Figure 3a, T6 heat treatment led to a higher increase in strength for A357 than a reinforcement with 35 vol.% SiC_p. An additional effect on the onset of plastic flow as well as the work hardening behavior can be expected by pore coarsening and grain coarsening resulting from the T6 heat treatment, see Section 3.1. However, compared to the precipitation hardening, only a minor effect of these microstructural changes can be assumed.

In the following, the investigated material is discussed compared to other particle-reinforced AMCs, taking into account their composition and processing methods: Sankhla et al. [42] also studied the compressive material behavior of AMCs manufactured by powder metallurgy. The composition with pure aluminum and the highest reinforcement content of 25 wt.% of SiC achieved a compression strength of about 260 MPa to 280 MPa. In the case of the sintered material investigated in our study, these values are not yet reached at the onset of yield (OYS), but they are clearly exceeded at a plastic deformation of about 2%. At a plastic deformation of 15%, the strength is higher by about one quarter than that of the pure Al/SiC_p composition. This is mainly due to the higher reinforcement content and alloying elements of our material, which lead to additional strengthening. The compression strength of a stir-cast A356 with a content of 6 wt.% SiC in the T6 heat-treated condition was 334.28 MPa, as investigated by Sunitha et al. [43]. This value is comparable to $\sigma_{15\%}$ of the as-sintered condition of our investigated AMC (Figure 3b), but the value for the T6 condition is even higher by about 70%. Again, the reinforcement content has a significant influence on the strength of the AMC. However, despite the higher volume fraction of age-hardenable aluminum and thus a greater potential amount of strengthening precipitates for the AMC of Sunitha et al. [43], the strength is significant lower. This seems to be contrary considering the strength that can be achieved by the T6 heat treatment (see Figure 3a), even without a reinforcement phase. It should, however, be taken into account, that both materials also differ in terms of Mg content and the manufacturing process which caused a coarser microstructure for the cast AMC [43].

A different ceramic reinforcement was used by Shao et al. [44]. They investigated a cold-sprayed composite with a pure aluminum matrix reinforced with 46 wt.% of Al_2O_3 ($\approx$40 vol.%). The maximum OYS achieved for this material was 317.4 MPa. The reinforcement content of the Al_2O_3/Al composite was higher by about 5 vol.% and the porosity was about ten times lower compared to our investigated material $SiC_p/A357$, which increased the OYS considerably. In addition, the capacity for work hardening was significantly reduced, resulting from an enhanced grain boundary strengthening effect due to the very fine (sub-)grain size of 1 μm or less [44] in the investigated condition that was gained by cold-spray additive manufacturing. As a consequence of the limited plasticity, the maximum strength of the 46 wt.% Al_2O_3/Al was already achieved before a compression of about 2%. In contrast, the grain size of the same material annealed at 500 °C was much coarser (approx. 7 μm on average) and lattice distortion was reduced [44]. Consequently, this heat treatment results in a general reduction in strength but increases the capacity for work hardening during compression. The initial compressive behavior of the annealed AMC is therefore more comparable to the investigated AMC with a similar grain size (see Figure 2). In summary, it can be concluded that not only the AMC composition, but also the manufacturing process and heat treatment affect the mechanical properties significantly.

3.3. *Compressive Behavior at 350 °C*

The results of the quasi-static compression tests conducted at 350 °C are given in Figure 4. Compared to the tests conducted at RT (Figure 3), the onset of yield (see Figure 4b) is at a significantly lower level of stress, i.e., the OYS of the as-sintered A357 is decreased by about 30% and for the as-sintered $SiC_p/A357$, the reduction is about 50%. This effect is more pronounced for the T6 heat-treated conditions which exhibit a decrease of about 60% and 70% for A357 and $SiC_p/A357$, respectively. This considerable reduction in strength arises on the one hand from the lower flow stress at elevated temperatures due to additional thermal activation of dislocation slip and on the other hand from static over-ageing of the T6 tempered samples during heating of the compression samples prior to testing.

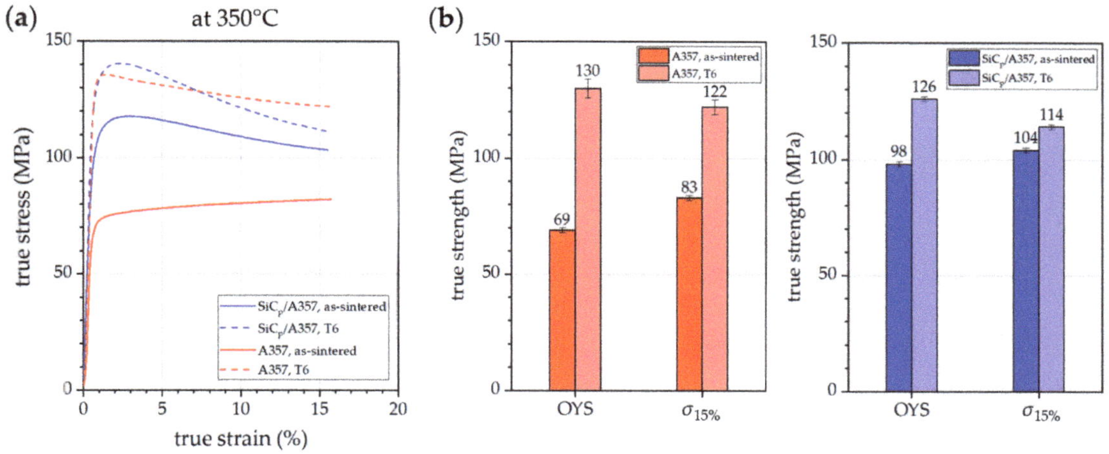

Figure 4. (**a**) Representative true stress–strain curves plotted up to a true compression of 15% and (**b**) mechanical properties obtained from compression tests at 350 °C for the as-sintered and T6 conditions.

As shown in the DSC curve of the T6 heat-treated condition of A357 (Figure 5, dashed red line), heating of this sample results in an endothermic peak at temperatures of about 210 °C to 270 °C and a subsequent pronounced exothermic peak with a maximum at about 310 °C followed by an overlapping peak at about 360 °C. These peaks can be assigned to the dissolution of initially present β″ precipitates (II) and the formation of β′ precipitates (IV) and β precipitates (V), respectively. As a consequence, these larger β′ (coarser, semi-coherent) and β (coarse, incoherent) precipitates hinder the dislocation movement to a lesser extent than the β″ precipitates which results in the lower stress that is necessary for the onset of plastic deformation compared to compression at RT.

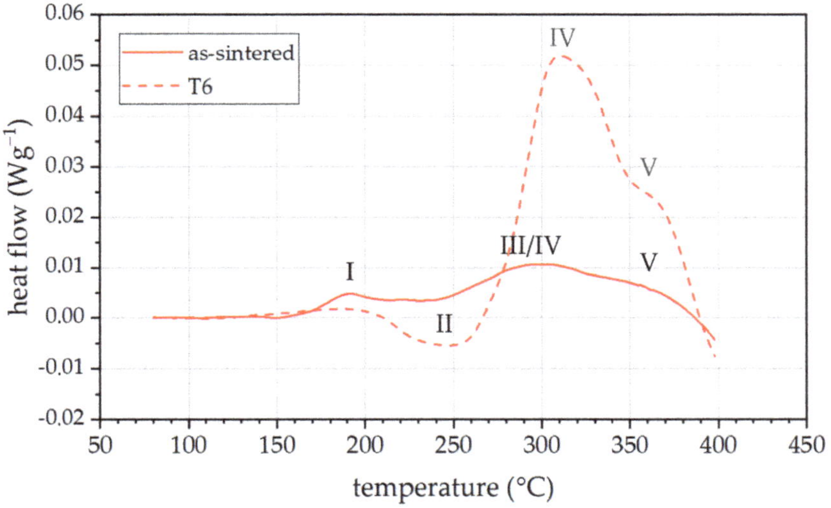

Figure 5. Differential scanning calorimetry (DSC) thermograms of the matrix alloy A357 in the as-sintered condition (solid red line) and T6 condition (dashed red line) obtained at a heating rate of 20 K/min.

In contrast, the DSC curve of the as-sintered condition (solid red line, Figure 5) exhibits only weak, exothermic peaks. This condition is characterized by the temperature progression during the sintering process, i.e., heating and holding at a temperature of 520 °C which can be considered as a suitable temperature for solution annealing and subsequent cooling to RT. Since the cooling rate in this process is not comparable to water-quenching, it can be expected that the atoms that were dissolved during sintering did not remain in the solid solution but already formed precipitates of different stages of the precipitation sequence during cooling. During the subsequent heating (e.g., prior to hot compression testing), these (caused by the low cooling rate) sparsely distributed precipitates grow while their coherency decreases. However, it can be assumed that the precipitation kinetic during testing at 350 °C of the as-sintered conditions starts with the formation of GP I zones at about 190 °C, see Figure 5 (I), which indicates that Si-, Mg- and Mg-Si co-clusters were already present after sintering [9]. This peak is followed by the overlapping formation of β'' (III), β' (IV) and β (V) precipitates, from about 240 °C to 380 °C. At about 380 °C, the precipitation ends and the dissolution of the Mg_2Si phases begins for both conditions. During testing, the temperature is at a level of the formation of the β precipitates (V).

Upon further plastic straining, the stress–strain curves (see Figure 3a) show different trends for the investigated material compositions and conditions. While the strength of the as-sintered A357 is steadily increasing, the other samples exhibit a material behavior typically observed for hot compression [30], i.e., a maximum of strength with a subsequent decrease in the compression stress. This results from additional softening effects initiated at elevated temperatures, namely DRV and DRX, which counteract the strengthening effect of work hardening. Furthermore, the effects of DP need to be considered, which affect the deformation behavior of the as-sintered and T6 heat-treated condition differently, as already discussed in terms of static precipitation during the heating prior to testing.

It can be assumed that the reinforcement and the present precipitations (size and distribution) influence DRV and DRX in a different way for the investigated materials. Regarding the reinforcement, it seems that DRV and DRX are more pronounced for SiC_p/A357 compared to A357. In general, it is assumed that the reinforcing particles enhance the multiplication and interaction of dislocations during hot deformation and thus promote dynamic recrystallisation [45].

An additional effect on the softening behavior seems to occur for the T6 conditions of both materials. Based on the model of Chen et al. [25], who investigated the hot deformation behavior of a hybrid AMC (micro-SiC and nano-TiB_2), a higher amount of dislocations accumulated at the nano-scaled particles compared to the micro-scaled particles. As a result, these small (incoherent) phases provide a greater driving force for DRX [25]. It can therefore be concluded that (nano-scaled) precipitates also enforce the formation of new grains due to a more continuously increasing pile up and tangle of dislocations [46].

Contrarily, the recrystallisation maps (see Figure 6) of specimens tested up to a deformation of 15% clearly show a higher content of recrystallized grains (blue) for the as-sintered A357, while the corresponding T6 condition is mainly characterized by deformed grains (red) with a higher orientation spread within the grains. This indicates that for the T6 conditions, DRX as well as DRV (low number of yellow, i.e., recovered grains) is inhibited due to a larger number of more regularly distributed precipitates compared to the as-sintered conditions where the spacing between precipitates is expected to be larger. The more regularly dispersed precipitates effectively pin dislocations and promote the dense arrangement of dislocations resulting in a pronounced gradient of misorientation within the grains. Chen et al. [47] furthermore found that the dispersed nanoparticles acted as obstacles for dislocation glide, cross slip and probably climb as well as the migration of low angle grain boundaries and therefore strongly impeded the recovery of the substructures. This confirms the observations in Figure 6b, where only a few small recrystallized grains appear along grain boundaries. The residual microstructure (grain boundaries and substructures) is probably conserved by the network of incoherent precipitates. However, a decrease in strength was noted. On this basis, it can be assumed that the further over-ageing and coarsening of the precipitates (Ostwald ripening) that was accelerated by the deformation (enhanced dynamic processes) led to the observed softening of the material.

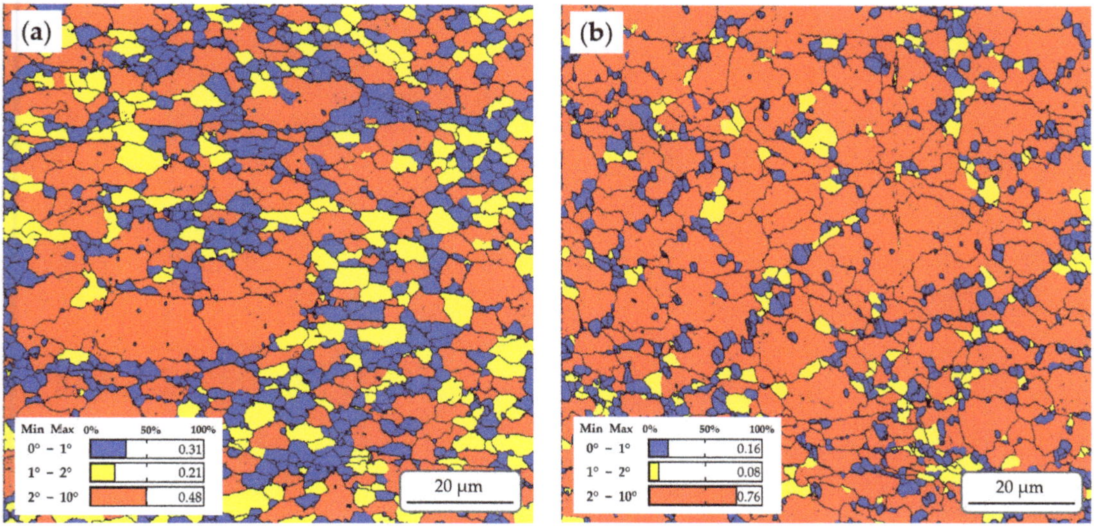

Figure 6. Recrystallisation maps obtained by EBSD measurements on the longitudinal sections of tested (compression ~ 15%) samples for (**a**) the as-sintered condition and (**b**) the T6 heat-treated condition of A357. High-angle grain boundaries (>15°) are marked by black lines.

In contrast, the as-sintered condition shows a balance between deformed (red) and recovered or recrystallized (yellow and blue) grains. In this case, the softening effects due to DRV and DRX are more pronounced in comparison to the T6 condition. However, the work hardening of the as-sintered condition is not as strongly reduced as expected. For instance, Liang and Zhang [48], who investigated the flow curves of an Al-Si-Mg alloy similar to A357 at different strain rates and temperatures, found that the flow stress at later stages of deformation still increases when there is insufficient time for stress dissipation (energy accumulation and dislocation annihilation), usually found at higher strain rates. According to the flow curves given in [48,49], there is a peak behavior for compression at 350 °C and at strain rates up to 0.1 s^{-1}. At a strain rate of 1 s^{-1} and higher, a plateau is observed, which is attributed to a balance between work hardening and dynamic softening. This indicates an additional strengthening effect for the as-sintered A357 in comparison to the T6 condition. Wu et al. [50] found that during hot extrusion the amount of Si particles of cast Al-Si alloys strongly affects the recrystallisation behavior. It is described that with higher Si particles density, more DRX is induced [50]. The Si content of the A357 alloy investigated in our study was not changed, but the density in terms of inter-particle spacing is higher for the as-sintered condition. In addition, the hot compression curves at 350 °C, at the lowest strain rate (10^{-2} s^{-1}) determined by Wu et al. [51], show that the smelting-cast near-eutectic Al-Si alloy with finer eutectic silicon has a less pronounced peak behavior than that with coarser Si particles. On the one hand, it can therefore be assumed that the mechanical behavior of the as-sintered condition results from the finer eutectic Si phase distribution (see Figure 1). On the other hand, this is due to the progressive new formation of (strengthening) precipitates, which in turn simultaneously favor the continuous growth of new grains (Figure 6). The rate of strengthening during the compression test continuously decreases (Figure 4a), which is presumably caused by the over-ageing of these new precipitates. This would ultimately result in a softening effect as observed for the T6 condition.

The mechanisms identified for the unreinforced material can presumably be applied to the AMCs. However, the extent of DRX is assumed to be significantly more pronounced due to the reinforcement [45], resulting in the observed stronger softening. Beside this, an additional reduction in strength was identified by Liu et al. [52]. They found that the reinforcement particles roll and slide within the matrix. Due to this, the particle skeleton becomes more unstable and the deformation behavior is dominated by the matrix [52]. This phenomenon leads to a higher plasticity for the whole composite material compared to deformation at RT. As a consequence, Graf et al. [53] observed an alignment of the reinforcement phase, i.e., an arrangement of the particles corresponding to the direction of material flow. Based on this, the evolution of the microstructure of the AMCs was analyzed and is discussed in detail in Section 3.4.

3.4. Microstructural Evolution of AMCs during Compression

Figure 7 presents optical micrographs of selected specimens tested up to a final compression of 50%. As indicated by the dashed orange ellipses, porosity is slightly reduced but still apparent in the center of the specimen tested at RT, while it is significantly reduced after testing at 350 °C. On the one hand, the high testing temperature led to a reduced flow stress due to thermal activation of dislocation movement and thus enhanced recovery and recrystallisation processes. On the other hand, the strength of the matrix of the T6 condition was additionally reduced by the coarser eutectic Si phase (caused by heat treatment [2]) and over-ageing (reduction in the number of, as well as increase in the spaces between, obstacles for dislocation movement). As a result, lower stresses were necessary to deform the specimen with increasing deformation. For both AMC conditions tested at 350 °C, the final porosity at the specimens' centers, determined by grey threshold values from optical microscopic images, was about 1% ± 0.5%. Compared to the initial porosity (see Table 3), the reduction in the residual pores is thus more pronounced for the T6 specimens than for the as-sintered specimens tested at 350 °C.

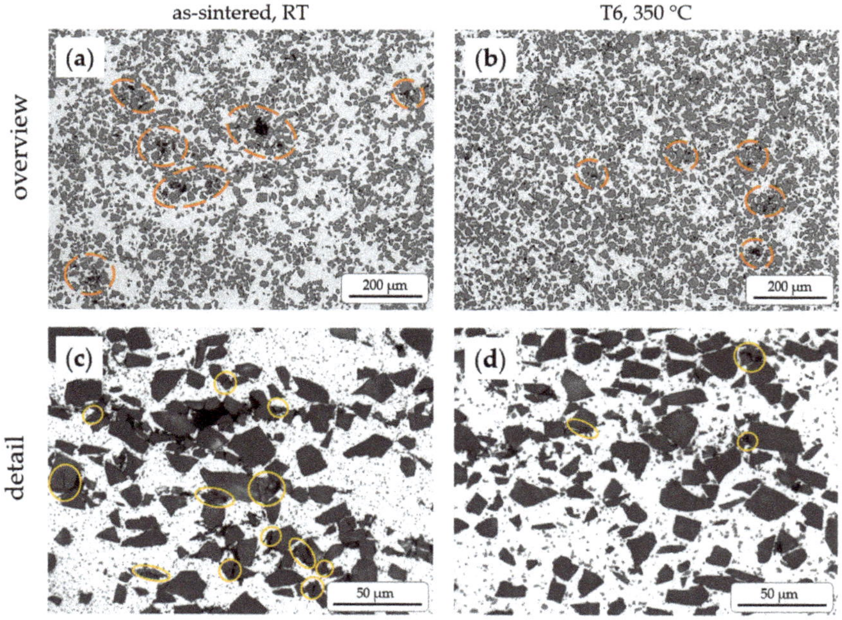

Figure 7. Optical microscopic images of the centers of AMC specimen tested up to a compression of 50% are given as overview (**a**,**b**) with the cross-sectional area perpendicular to the compression stress, and in detail (**c**,**d**) with the cross-sectional area parallel to the direction of the compression stress. The dashed orange ellipses mark residual porosity and the solid yellow ellipses mark fractured SiC_p which are not incorporated in or surrounded by the aluminum matrix.

In comparison with Figure 1, it also becomes clear that the specimens tested at RT show a similar arrangement of SiC_p, while the distribution of SiC_p became more regular after testing at 350 °C for both conditions. Example microstructures are given in Figure 7a,b. In the specimen tested at RT, there are still local areas with more or less accumulated SiC_p caused by the former powder mixture prior to the consolidation process. An opposite effect can be observed for the specimen tested at 350 °C. According to Liu et al. [52], the ductile matrix led to an easy sliding and rolling of the particles. In conclusion, this confirms the assumption of a rearrangement favorable for deformation and perhaps also the fast incorporation or surrounding of fractured (see T6 condition) reinforcement particles in or by the matrix during compression.

On a closer look, the qualitative microstructural analysis (see details in Figure 7) and the evaluation of the particle size distribution (see Figure 8) show that during testing at RT the fracture of SiC particles is highest: the contents of particle fractions between IV and VII (from ≥12.5 μm to <20 μm) are significantly reduced and the amount of particles of smaller fractions (I and II, from ≥5 μm to <10 μm) increased compared to the initial condition (Figure 8a,c). To a lesser extent, this can also be observed for the T6 condition tested at 350 °C (Figure 8d). Chen et al. [24] and Chen et al. [25] also observed fracturing of the ceramic particles when the stress concentration was higher than the fracture strength of the SiC_p. It is clear that at lower temperatures (here: RT) the stress compensation of the matrix is reduced compared to higher temperatures (here: 350 °C), when softening effects, like DRV and DRX, can take place. The probability of damaging the SiC particles is therefore increased for the specimens tested at RT. Unlike the other three conditions, the SiC particle size fractions of the as-sintered condition tested at 350 °C is nearly unchanged compared to the initial condition. As found in Section 3.3., different or reduced softening effects took place in the matrix during compression. It was observed that for the as-sintered condition, DRV and DRX were less inhibited due to the absence of a finely distributed precipitation network in comparison to the T6 condition. Therefore, high concentrations of piled-up dislocations could be compensated. As a consequence, nearly no particle fracturing stress concentration occurred during further compression.

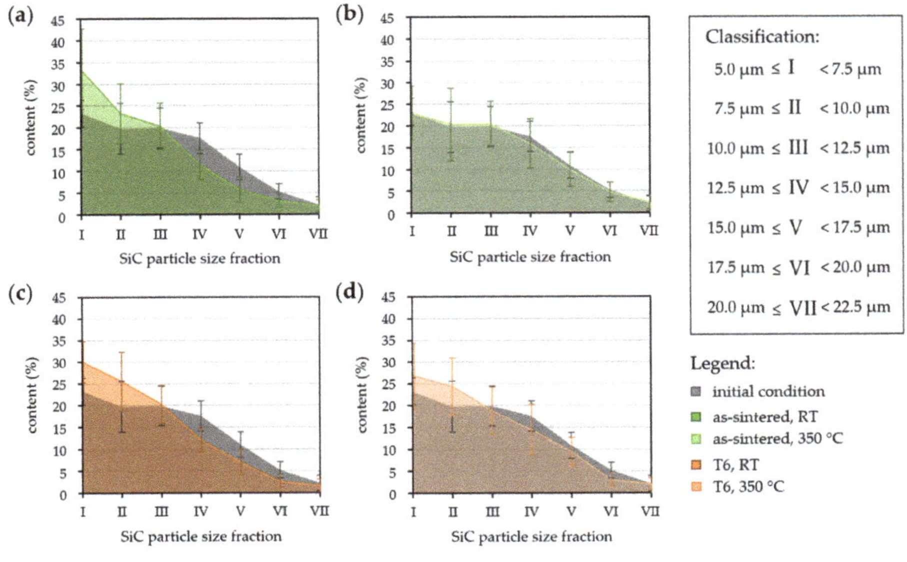

Figure 8. Variation in the SiC particle size distribution before (initial, grey) and after compression tests for the as-sintered ((**a**)—tested at RT, (**b**)—tested at 350 °C) and T6 heat-treated condition ((**c**)—tested at RT, (**d**)—tested at 350 °C) after a compression of 50%.

It should be noted that the content of the SiC particle fraction III (from ≥10 μm to <12.5 μm) is constant before and after testing for all conditions and may therefore be a critical size for the quasi-static compressive deformation of the investigated AMC composition. However, it cannot be clearly separated whether particles of this size are a maximum size for no particle damage or whether particles of this size were formerly a part of bigger particles.

4. Summary

In this study, the compressive behavior of a sintered A357 cast aluminum powder with and without a ceramic particle reinforcement (35 vol.% SiC_p, $d_{50} \approx 21$ μm) was investigated for the as-sintered conditions without subsequent heat treatment as well as for the T6 heat-treated conditions at RT and at 350 °C. The main results are as follows:

- *Testing at RT:*

In general, the addition of reinforcing particles as well as the T6 heat treatment result in an expected increase in strength. However, the strengthening effect of the T6 heat treatment was more pronounced than the effect of particle reinforcement. While the reinforcement led to a doubling of the OYS from about 100 MPa to 200 MPa, the T6 condition led to an increase in OYS by 200 MPa for both materials.

- *Testing at 350 °C:*

Due to softening effects, the strength of all material conditions decreased significantly compared to the RT compression tests. Depending on the initial microstructure (reinforcement particles and precipitates), thermally activated processes like DRV, DRX and DP affect the compressive behavior to varying degrees. It was found that reinforcement generally promotes softening effects, resulting in a pronounced peak behavior in the stress–strain curve. The T6 heat treatment enhances this softening effect by dynamic over-ageing of the already present precipitates. However, for the as-sintered conditions, DP is likely to counteract the softening effects by formation of new precipitates.

In addition, it was observed that SiC particles were fractured during a compression of up to about 50% at RT. At a testing temperature of 350 °C, this effect of particle fracturing is reduced (T6) or even undetectable (as-sintered) which can be attributed to an easier rolling and sliding of the particles within the more ductile aluminum matrix. This additionally led to a rearrangement of the particles.

This more regular distribution of the SiC_p provides the potential for an improved further formability and probably more homogenous properties of the composite with regard to hot forming processes. Furthermore, with regard to wear applications, for which highly ceramic particle-reinforced AMCs are predestined, the more regularly distributed SiC_p could additionally reduce the wear of this material. The potential and significance for processing and application need to be verified by investigations with regard to different application conditions.

Author Contributions: Conceptualization, S.J.H.; methodology, S.J.H.; validation, S.J.H. and N.B.; formal analysis, S.J.H. and N.B.; investigation, S.J.H. and N.B.; resources, T.L.; data curation, S.J.H. and N.B.; writing—original draft preparation, S.J.H. and N.B.; writing—review and editing, T.G. and T.L.; visualization, S.J.H.; supervision, T.G and T.L. All authors have read and agreed to the published version of the manuscript.

Funding: This research received no external funding.

Data Availability Statement: The authors confirm that the data to support the findings of this study are available within the article.

Acknowledgments: The authors would like to thank Felix Schubert, Steffen Clauß, Mario Scholze and Siddhartha Biswas for their help.

Conflicts of Interest: The authors declare no conflicts of interest.

References

1. Lakshmikanthan, A.; Prabhu, T.R.; Udayagiri, S.B.; Koppad, P.G.; Gupta, M.; Munishamaiah, K.; Bontha, S. The Effect of Heat Treatment on the Mechanical and Tribological Properties of Dual Size SiC Reinforced A357 Matrix Composites. *J. Mater. Res. Technol.* **2020**, *9*, 6434–6452. [CrossRef]
2. Hirsch, S.J.; Winter, L.; Grund, T.; Lampke, T. Heat Treatment Influencing Porosity and Tensile Properties of Field Assisted Sintered AlSi7Mg0.6. *Materials* **2022**, *15*, 2503. [CrossRef] [PubMed]
3. Wang, Q.G. Microstructural Effects on the Tensile and Fracture Behavior of Aluminum Casting Alloys A356/357. *Met. Mater. Trans. A* **2003**, *34*, 2887–2899. [CrossRef]
4. Aversa, A.; Lorusso, M.; Trevisan, F.; Ambrosio, E.P.; Calignano, F.; Manfredi, D.; Biamino, S.; Fino, P.; Lombardi, M.; Pavese, M. Effect of Process and Post-Process Conditions on the Mechanical Properties of an A357 Alloy Produced via Laser Powder Bed Fusion. *Metals* **2017**, *7*, 68. [CrossRef]
5. Zhang, D.; Sun, S.; Qiu, D.; Gibson, M.A.; Dargusch, M.S.; Brandt, M.; Qian, M.; Easton, M. Metal Alloys for Fusion-Based Additive Manufacturing. *Adv. Eng. Mater.* **2018**, *20*, 1700952. [CrossRef]
6. Lathabai, S. Chapter 2—Additive Manufacturing of Aluminium-Based Alloys and Composites. In *Fundamentals of Aluminium Metallurgy*; Lumley, R.N., Ed.; Woodhead Publishing Series in Metals and Surface Engineering; Woodhead Publishing: Duxford, UK, 2018; pp. 47–92. ISBN 978-0-08-102063-0.
7. Zandbergen, M.W.; Cerezo, A.; Smith, G.D.W. Study of Precipitation in Al–Mg–Si Alloys by Atom Probe Tomography II. Influence of Cu Additions. *Acta Mater.* **2015**, *101*, 149–158. [CrossRef]
8. Pogatscher, S.; Antrekowitsch, H.; Leitner, H.; Ebner, T.; Uggowitzer, P.J. Mechanisms Controlling the Artificial Aging of Al–Mg–Si Alloys. *Acta Mater.* **2011**, *59*, 3352–3363. [CrossRef]
9. Winter, L.; Hockauf, K.; Scholze, M.; Hellmig, R.J.; Lampke, T. Influence of Pre-Aging on the Artificial Aging Behavior of a 6056 Aluminum Alloy after Conventional Extrusion. *Metals* **2021**, *11*, 385. [CrossRef]
10. Adebisi, A.A.; Maleque, M.A.; Rahman, M.M. Metal Matrix Composite Brake Rotor: Historical Development and Product Life Cycle Analysis. *Int. J. Automot. Mech. Eng.* **2011**, *4*, 471–480. [CrossRef]
11. Natarajan, N.; Vijayarangan, S.; Rajendran, I. Wear Behaviour of A356/25SiCp Aluminium Matrix Composites Sliding against Automobile Friction Material. *Wear* **2006**, *261*, 812–822. [CrossRef]
12. Daoud, A.; Abou El-khair, M.T. Wear and Friction Behavior of Sand Cast Brake Rotor Made of A359-20vol% SiC Particle Composites Sliding against Automobile Friction Material. *Tribol. Int.* **2010**, *43*, 544–553. [CrossRef]
13. Joshi, Y.G.; Gupta, A.R.; Shingarwade, R.U. Scrutinization of A356/25sicp AMC and Gray Cast Iron as Brake Rotor Material. *Int. J. Res. Advent Technol.* **2014**, *2*, 88–93.
14. Baig, M.M.A.; Al-Qutub, A.M.; Allam, I.M.; Patel, F.; Mohammed, A.S. Tribological Performance of Sub-Micron Al_2O_3-Reinforced Aluminum Composite Brake Rotor Material. *Arab. J. Sci. Eng.* **2021**, *46*, 2691–2700. [CrossRef]
15. Idusuyi, N.; Babajide, I.; Ajayi, O.; Olugasa, T. A Computational Study on the Use of an Aluminium Metal Matrix Composite and Aramid as Alternative Brake Disc and Brake Pad Material. *J. Eng.* **2014**, *2014*, 494697. [CrossRef]
16. Nakanishi, H.; Kakihara, K.; Nakayama, A.; Murayama, T. Development of Aluminum Metal Matrix Composites (Al-MMC) Brake Rotor and Pad. *JSAE Rev.* **2002**, *23*, 365–370. [CrossRef]
17. Chawla, N.; Chawla, K.K. Metal Matrix Composites; Springer New York, US, 2006; ISBN 978-0-387-28567-2.
18. Singh, B.; Kumar, I.; Saxena, K.K.; Mohammed, K.A.; Ijaz Khan, M.; Ben Moussa, S.; Shukhratovich Abdullaev, S. A Future Prospects and Current Scenario of Aluminium Metal Matrix Composites Characteristics. *Alex. Eng. J.* **2023**, *76*, 1–17. [CrossRef]
19. Kumar, P.K.D.; Gnanaraj, S.D. Aluminium-Silicon Based Metal Matrix Composites for Brake Rotor Applications: A Review. *Eng. Res. Express* **2023**, *5*, 022002. [CrossRef]
20. Mondal, D.P.; Ganesh, N.V.; Muneshwar, V.S.; Das, S.; Ramakrishnan, N. Effect of SiC Concentration and Strain Rate on the Compressive Deformation Behaviour of 2014Al-SiCp Composite. *Mater. Sci. Eng. A* **2006**, *433*, 18–31. [CrossRef]
21. Ye, T.; Xu, Y.; Ren, J. Effects of SiC Particle Size on Mechanical Properties of SiC Particle Reinforced Aluminum Metal Matrix Composite. *Mater. Sci. Eng. A* **2019**, *753*, 146–155. [CrossRef]
22. Sajjadi, S.A.; Ezatpour, H.R.; Beygi, H. Microstructure and Mechanical Properties of Al–Al_2O_3 Micro and Nano Composites Fabricated by Stir Casting. *Mater. Sci. Eng. A* **2011**, *528*, 8765–8771. [CrossRef]
23. Abouelmagd, G. Hot Deformation and Wear Resistance of P/M Aluminium Metal Matrix Composites. *J. Mater. Process. Technol.* **2004**, *155–156*, 1395–1401. [CrossRef]
24. Chen, S.; Teng, J.; Luo, H.; Wang, Y.; Zhang, H. Hot Deformation Characteristics and Mechanism of PM 8009Al/SiC Particle Reinforced Composites. *Mater. Sci. Eng. A* **2017**, *697*, 194–202. [CrossRef]
25. Chen, X.; Fu, D.; Teng, J.; Zhang, H. Hot Deformation Behavior and Mechanism of Hybrid Aluminum-Matrix Composites Reinforced with Micro-SiC and Nano-TiB2. *J. Alloys Compd.* **2018**, *753*, 566–575. [CrossRef]
26. Hao, S.; Xie, J.; Wang, A.; Wang, W.; Li, J.; Sun, H. Hot Deformation Behaviors of 35%SiCp/2024Al Metal Matrix Composites. *Trans. Nonferrous Met. Soc. China* **2014**, *24*, 2468–2474. [CrossRef]
27. Hao, S.M.; Xie, J.P. Hot Deformation Behaviors of SiCp/2024 Aluminum Composites. *Adv. Mater. Res.* **2014**, *833*, 271–275. [CrossRef]
28. Kai, X.; Zhao, Y.; Wang, A.; Wang, C.; Mao, Z. Hot Deformation Behavior of in Situ Nano ZrB2 Reinforced 2024Al Matrix Composite. *Compos. Sci. Technol.* **2015**, *116*, 1–8. [CrossRef]

29. Khodabakhshi, F.; Gerlich, A.P.; Simchi, A.; Kokabi, A.H. Hot Deformation Behavior of an Aluminum-Matrix Hybrid Nanocomposite Fabricated by Friction Stir Processing. *Mater. Sci. Eng. A* **2015**, *626*, 458–466. [CrossRef]
30. McQueen, H.J. Development of Dynamic Recrystallization Theory. *Mater. Sci. Eng. A* **2004**, *387–389*, 203–208. [CrossRef]
31. Asgharzadeh, H.; Simchi, A.; Kim, H.S. High-Temperature Deformation and Structural Restoration of a Nanostructured Al Alloy. *Scr. Mater.* **2012**, *66*, 911–914. [CrossRef]
32. Fan, X.H.; Li, M.; Li, D.Y.; Shao, Y.C.; Zhang, S.R.; Peng, Y.H. Dynamic Recrystallisation and Dynamic Precipitation in AA6061 Aluminium Alloy during Hot Deformation. *Mater. Sci. Technol.* **2014**, *30*, 1263–1272. [CrossRef]
33. Roven, H.J.; Liu, M.; Werenskiold, J.C. Dynamic Precipitation during Severe Plastic Deformation of an Al–Mg–Si Aluminium Alloy. *Mater. Sci. Eng. A* **2008**, *483–484*, 54–58. [CrossRef]
34. Teichmann, K.; Marioara, C.D.; Pedersen, K.O.; Marthinsen, K. The Effect of Simultaneous Deformation and Annealing on the Precipitation Behaviour and Mechanical Properties of an Al–Mg–Si Alloy. *Mater. Sci. Eng. A* **2013**, *565*, 228–235. [CrossRef]
35. *DIN EN 1706*; Aluminium and Aluminium Alloys—Castings—Chemical Composition and Mechanical Properties. Deutsches Institut fur Nor-mung: Berlin, Germany, 2021.
36. Hirsch, S.J.; Grund, T.; Lampke, T. Towards Closed-Loop Recycling of Ceramic Particle-Reinforced Aluminium Alloys: Comparative Study of Resistance-Heating Sintered Primary and Solid-State Recycled Secondary SiCp/AlSi7Mg Composites. *Crystals* **2023**, *13*, 830. [CrossRef]
37. Pippig, R.; Hirsch, S.J.; Grund, T.; Lampke, T. Influence of Metal Matrix Powder Size on the Tensile Strength of a SiCp/AlSi7Mg0,6 Composite Produced by Field Assisted Sintering Technique. *IOP Conf. Ser. Mater. Sci. Eng.* **2021**, *1147*, 012020. [CrossRef]
38. Fan, H.; Witvrouw, A.; Wolf-Monheim, F.; Souschek, R.; Yang, S. Effects of Substrate Surface Treatments on Hybrid Manufacturing of AlSi7Mg Using Die Casting and Selective Laser Melting. *J. Mater. Sci. Technol.* **2023**, *156*, 142–156. [CrossRef]
39. Yadav, P.; Rigo, O.; Arvieu, C.; Lacoste, E. Microstructural and Mechanical Aspects of AlSi7Mg0.6 Alloy Related to Scanning Strategies in L-PBF. *Int. J. Adv. Manuf. Technol.* **2022**, *120*, 6205–6223. [CrossRef]
40. Ming, X.; Song, D.; Yu, A.; Tan, H.; Zhang, Q.; Zhang, Z.; Chen, J.; Lin, X. Effect of Heat Treatment on Microstructure, Mechanical and Thermal Properties of Selective Laser Melted AlSi7Mg Alloy. *J. Alloys Compd.* **2023**, *945*, 169278. [CrossRef]
41. Hwang, W.J.; Bang, G.B.; Choa, S.-H. Effect of a Stress Relief Heat Treatment of AlSi7Mg and AlSi10Mg Alloys on Mechanical and Electrical Properties According to Silicon Precipitation. *Met. Mater. Int.* **2023**, *29*, 1311–1322. [CrossRef]
42. Sankhla, A.M.; Patel, K.M.; Makhesana, M.A.; Giasin, K.; Pimenov, D.Y.; Wojciechowski, S.; Khanna, N. Effect of Mixing Method and Particle Size on Hardness and Compressive Strength of Aluminium Based Metal Matrix Composite Prepared through Powder Metallurgy Route. *J. Mater. Res. Technol.* **2022**, *18*, 282–292. [CrossRef]
43. Sunitha, K.; Gurusami, K.; Rajeswari, N.; Kathikraja, A.; Saravanan, L. Investigation of Mechanical Properties of SiC Particle Reinforced Aluminium A356/LM25 Alloy Composite. *Mater. Today Proc.* **2022**, *66*, 1983–1987. [CrossRef]
44. Shao, C.; Lo, C.; Bhagavathula, K.; McDonald, A.; Hogan, J. High Strength Particulate Aluminum Matrix Composite Design: Synergistic Strengthening Strategy. *Compos. Commun.* **2021**, *25*, 100697. [CrossRef]
45. Liu, T.-S.; Dong, B.-X.; Yang, H.-Y.; Qiu, F.; Shu, S.-L.; Jiang, Q.-C. Review on Role of Intermetallic and Ceramic Particles in Recrystallization Driving Force and Microstructure of Wrought Al Alloys. *J. Mater. Res. Technol.* **2023**, *27*, 3374–3395. [CrossRef]
46. She, H.; Shu, D.; Dong, A.; Wang, J.; Sun, B.; Lai, H. Relationship of Particle Stimulated Nucleation, Recrystallization and Mechanical Properties Responding to Fe and Si Contents in Hot-Extruded 7055 Aluminum Alloys. *J. Mater. Sci. Technol.* **2019**, *35*, 2570–2581. [CrossRef]
47. Chen, Z.; Sun, G.A.; Wu, Y.; Mathon, M.H.; Borbely, A.; Chen, D.; Ji, G.; Wang, M.L.; Zhong, S.Y.; Wang, H.W. Multi-Scale Study of Microstructure Evolution in Hot Extruded Nano-Sized TiB2 Particle Reinforced Aluminum Composites. *Mater. Des.* **2017**, *116*, 577–590. [CrossRef]
48. Liang, Z.; Zhang, Q. Quasi-Static Loading Responses and Constitutive Modeling of Al–Si–Mg Alloy. *Metals* **2018**, *8*, 838. [CrossRef]
49. Kang, F.; Wei, S.; Zhang, J.; Wang, E.; Fan, D.; Wang, S. Hot Processing Maps and Microstructural Characteristics of A357 Alloy. *J. Mater. Eng. Perform.* **2020**, *29*, 7352–7360. [CrossRef]
50. Wu, Y.; Liao, H.; Yang, J.; Zhou, K. Effect of Si Content on Dynamic Recrystallization of Al–Si–Mg Alloys During Hot Extrusion. *J. Mater. Sci. Technol.* **2014**, *30*, 1271–1277. [CrossRef]
51. Wu, Y.; Liu, C.; Liao, H.; Jiang, J.; Ma, A. Joint Effect of Micro-Sized Si Particles and Nano-Sized Dispersoids on the Flow Behavior and Dynamic Recrystallization of near-Eutectic Al–Si Based Alloys during Hot Compression. *J. Alloys Compd.* **2021**, *856*, 158072. [CrossRef]
52. Liu, B.; Huang, W.; Wang, H.; Wang, M.; Li, X. Compressive Behavior of High Particle Content B4C/Al Composite at Elevated Temperature. *Trans. Nonferrous Met. Soc. China* **2013**, *23*, 2826–2832. [CrossRef]
53. Graf, M.; Pippig, R.; Lehnert, T.; Jedynak, A.; Härtel, S. Aluminium-Matrix-Composites (AMC) for Hot Forged Components. In Proceedings of the 14th International Conference on the Technology of Plasticity—Current Trends in the Technology of Plasticity; Mocellin, K., Bouchard, P.-O., Bigot, R., Balan, T., Eds.; Springer Nature Switzerland: Cham, Switzerland, 2024; pp. 36–47.

crystals

MDPI

Article

Aluminium-Silicon Lightweight Thermal Management Alloys with Controlled Thermal Expansion

Peter Lewis [1,*], Andrew Tarrant [1], Andreas Frehn [2], Fritz Grensing [3], James Nicholson [1], Nick Farrah [3] and Martyn Acreman [3]

[1] Materion UK Ltd., Farnborough GU14 6XE, UK
[2] Materion Brush GmbH, 70499 Stuttgart, Germany
[3] Materion Corporation, Mayfield Heights, OH 44124, USA
* Correspondence: peter.lewis@materion.com

Abstract: With the ever-growing emphasis on global decarbonization and rapid increases in the power densities of electronics equipment in recent years, new methods and lightweight materials have been developed to manage heat load as well as interfacial stresses associated with coefficient of thermal expansion (CTE) mismatches between components. The Al–Si system provides an attractive combination of CTE performance and high thermal conductivity whilst being a very lightweight option. Such materials are of interest to industries where thermal management is a key design criterion, such as the aerospace, automotive, consumer electronics, defense, EV, and space sectors. This paper will describe the development and manufacture of a family of high-performance hyper-eutectic Al–Si alloys (AyontEX™) by a powder metallurgy method. These alloys are of particular interest for structural heat sink applications that require high reliability under thermal cycling (CTE of 17 μm/(m·°C)), as well as reflective optics and instrument assemblies that require good thermal and mechanical stability (CTE of 13 μm/(m·°C)). Critical performance relationships are presented, coupled with the microstructural, physical, and mechanical properties of these Al–Si alloys.

Keywords: CTE; heatsink; HIP; hypereutectic; mechanical alloying; metal matrix composite; MMC; powder metallurgy; reflective optics; thermal stability

1. Introduction

With the ever-growing emphasis on global decarbonization and rapid increases in the power densities of electronics equipment in recent years, thermal management has become an escalating engineering challenge. Thus, emphasis on designing systems that allow heat to be transported, stored, or expelled in a manner that enables key system elements to operate with high reliability within a specific temperature range is essential [1]. This raises a need for new methods and lightweight materials to manage heat load as well as interfacial stresses associated with coefficient of thermal expansion (CTE) mismatches between components. Example materials include metal matrix composites (MMCs), such as aluminium–silicon carbide (Al–SiC) and beryllium-beryllium oxide (Be–BeO). However, metallic systems, such as aluminum–beryllium (Al–Be) and aluminum–silicon (Al–Si) alloys, can also be applicable. Each of these material systems presents a unique set of performance properties and manufacturing challenges.

The Al–Si system provides an attractive combination of tailorable CTE and high thermal conductivity, while supporting a lightweight design. Al–Si alloys are of interest for structural heat sink applications that require high reliability under thermal cycling, where a CTE of 17 μm/(m·°C) is desirable to minimize strain due to CTE mismatch between mating copper components. Reflective optics and instrument assemblies are also application areas, in which a CTE close to 13 μm/(m·°C) is necessary to match the nickel plating typically applied to such components whilst providing good thermal stability over broad operating temperature ranges [2–5].

The production of high Si Al-based alloys via traditional liquid-state processing routes can be problematic due to the challenges posed in achieving good control of the size and morphology of the primary Si phase [6]. Often this can mean that complex alloying additions or unique solidification processing methods are necessary to achieve the properties required for high-performance applications [7,8]. In contrast, manufacturing via a powder metallurgy route enables a wider range of compositions and can provide greater strength and homogeneous-microstructure-associated mechanical integrity [9,10]. Furthermore, powder metallurgy can provide near-net shape capability, which has been proven to lead to reduced waste generation and lower energy consumption per unit mass in comparison to traditional forming methods [11]. By processing in the solid-state, any possible detrimental diffusional or chemical reactions between constituent phases can be minimized. Avoiding molten processing allows the creation of non-equilibrium phase structures. This enables extensive opportunities for novel and interesting material combinations to enhance performance beyond traditional engineering materials.

Mechanical alloying is a solid-state mixing process in which MMC materials and metallic alloys are combined using powder metallurgy methods. As no melting is required, the reinforcement volume fraction and particle size are controlled by raw material selection. Thereby, this process enables precise control of reinforcement particle sizes with higher reinforcement volume fractions, both of which are critical in maximizing and optimizing the property benefits that can be achieved with MMC systems relative to the monolithic material [12]. Mechanical alloying was first developed in the 1960s with the aim of fabricating a nickel-based superalloy with homogeneous distribution of oxide reinforcements for gas turbine applications [13]. Since then, many material combinations have been explored [14–26]. The basic principle is the cyclical welding and fracture of ductile and brittle powders via repetitive impact, resulting in a distribution of reinforcement within the metallic matrix particles [27]. However, the optimization of key parameters is necessary to achieve a homogeneous and stable process [24–26,28,29]. This composite powder then acts as an input into downstream consolidation processes, such as hot isostatic pressing (HIP) or extrusion.

Materion's mechanical alloying and powder metallurgy technologies have been used for decades to produce commercially available particle reinforced aluminium–SiC SupremEX® MMCs [30–32]. It should be noted that the industrial scale of this technology is highly unique. The total capacity of Materion UK's mechanical alloying facility is 300 MT per annum. It is understood that light alloy mechanical alloying capability at this scale does not exist anywhere else in the world. A schematic summarizing this powder metallurgy route is given by Figure 1. These MMCs have been shown to display a very fine and homogeneous microstructure, with precise control of particle sizes [33]. The resultant composite materials are known to be isotropic with high modulus, strength, wear, and fatigue performance [30,34], but maintain the machining, fabrication, coating, and processing characteristics of conventional aluminium alloys.

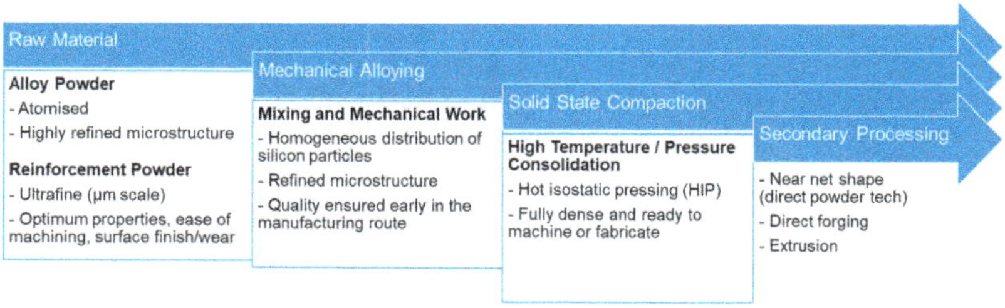

Figure 1. Schematic summarizing Materion's mechanical alloying powder metallurgy routes.

This paper will describe manufacturing of hypereutectic Al–Si alloys via a powder metallurgy route utilizing Materion's novel mechanical alloying process. Due to the flexibility of the manufacturing route, this lightweight material system has been finely tuned to achieve the specific and desirable CTE values referenced above. A detailed understanding of the relationship between chemistry and CTE has been developed. Critical performance relationships such as this will be presented, coupled with the microstructure and physical and mechanical properties of these Al–Si alloys.

2. Materials and Methods

The powder metallurgy processing described by Figure 1 and [30,31] was leveraged to combine elemental Si with Al alloy powders. The powder mixtures were mechanically alloyed until the Si was distributed within the Al matrix, with an overall refined and homogeneous microstructure. To achieve an optimal balance of mechanical and thermal properties, a 6063-alloy composition was selected, along with high-purity silicon powders. A series of samples with chemistry ranging from 15–55 wt% Si were processed. Initially the mechanically alloyed powder was loaded in a vibratory manner into an aluminium can, before degassing at an elevated temperature under vacuum conditions to eliminate air and any absorbed water vapor. Following this, the HIP process was applied to form fully dense cylindrical billets of material, approximately Ø100 mm × 170 mm in size. This compaction method is summarized by Figure 2.

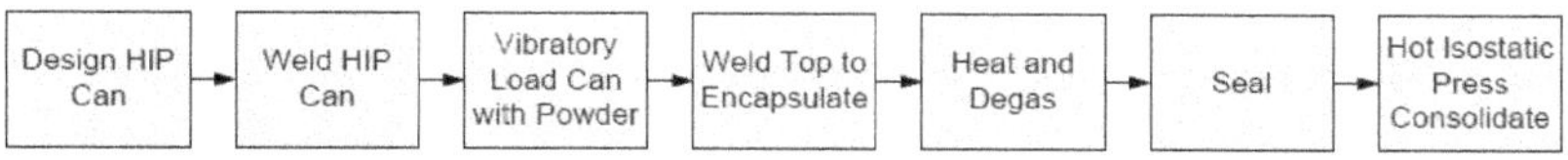

Figure 2. Major steps to HIP consolidate Al–Si powders made by the mechanical alloying process.

Samples for CTE evaluation were extracted from these billets and tested per ASTM E228-17 in order to understand the relationship between composition and this critical property. The samples were heat treated to a T6 condition prior to CTE measurement using the following method: solution treatment at 535 °C for 1 h, followed by a cold water (<30 °C) quench (CWQ), and finally, artificial aging at 175 °C for 2 h. All test samples were heat treated with a maximum section thickness less than or equal to 25 mm.

This initial CTE evaluation enabled derivation of the necessary compositions to achieve a CTE close to the target values of 17 μm/(m·°C) and 13 μm/(m·°C) at 28% and 42% Si, respectively. This is depicted graphically by Figure 3. Larger scale cylindrical billets (approx. Ø150 mm × 600 mm) and cuboid billets (approx. 180 mm × 180 mm × 300 mm) were subsequently manufactured at the derived compositions using the same manufacturing conditions to confirm the scalability of the process. These materials were then tested in detail to confirm their microstructure, room temperature tensile properties, density, and electrical conductivity. Tensile testing was carried out using an Instron 3369 loading frame with a 50 kN capacity. Multiple heat treatments were assessed, including the T6 CWQ heat treatment condition defined above, but also the same heat treatment using a 25% polymer-glycol quench (PGQ) medium, as well as the T1 (air-cooled from HIP) condition. Cylindrical specimens were used, with a 5 mm diameter and 25 mm gauge length. Test control was by a constant strain rate of 1.4×10^{-4} mm/mm/s, until failure. Microstructural analysis was completed using a Zeiss AxioLab5 optical microscope equipped with ZEN core imaging software. CTE was again evaluated per ASTM E228-17 but over a broader temperature range of −100 °C to 200 °C, whilst room temperature thermal diffusivity and specific heat capacity were measured per ASTM E1461-13 and ASTM E1269-11, respectively. This thermal testing was carried out for both the T6 PGQ and T1 heat treatment conditions described above, to assess sensitivity of these properties to heat treatment. Thermal conductivity was calculated as the product of thermal diffusivity, specific heat capacity, and density, per Equation (1) of ASTM E1461-13 [35]. Finally, a series of components representing example applications for each of the alloy components were produced, to assess manufacturability.

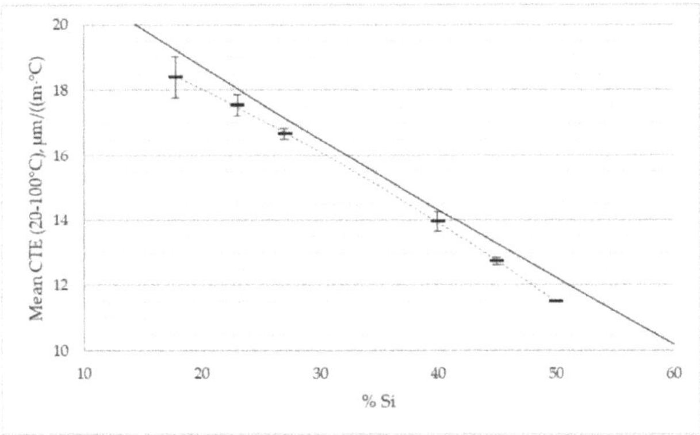

Figure 3. Effect of Si content on the CTE of the manufactured Al–Si alloys (dashed line) versus the simple rule of mixtures calculation (solid line). Data points are presented as mean values with 90% confidence intervals, where multiple tests were completed.

3. Results

3.1. Microstructure

An ultra-fine and homogeneous distribution of Si particles within the Al alloy was achieved for all Si contents tested, as Figure 4 shows. It can also be observed that the mechanically alloyed powder was generally blocky and irregular in shape across all compositions. This was consistent with past Al–SiC materials mechanically alloyed by the same process [30,31]. In the consolidated (post-HIP) form, the microstructure was seen to be consistent in all tested directions, indicating isotropic properties as expected. Typical optical micrographs of the 6063 + 42% Si alloy are presented Figure 5. Here, the ultrafine and homogeneous microstructure is again evident. The primary silicon size was observed to be an average of 3–4 µm in diameter, with almost all silicon particles within the 1–7 µm range.

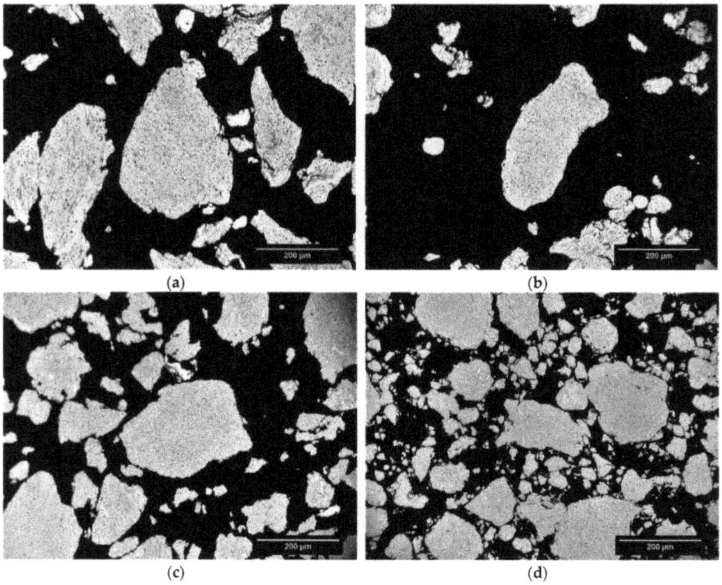

Figure 4. Mechanically alloyed Al–Si powders at (**a**) 18 wt% Si, (**b**) 27 wt% Si, (**c**) 40 wt% Si, and (**d**) 55 wt% Si (scale is intentionally withheld).

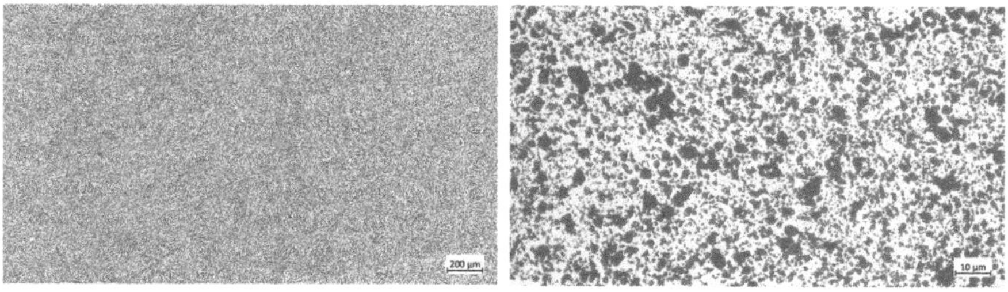

(**a**) (**b**)

Figure 5. AyontEX 13 HIP-processed billet optical micrographs at (**a**) 50× magnification and (**b**) 1000× magnification.

3.2. *Coefficient of Thermal Expansion*

The achieved CTE matches to commercially pure copper and nickel are summarized by Figure 6 via comparison to reference data from the literature [36–39]. This shows an exact mean CTE match to copper for the 20–50 °C temperature range. The 28% Si alloy was determined to have a marginally lower CTE for temperatures below room temperature and a marginally higher CTE at temperatures greater than 50 °C. The mean CTE value at all temperatures, as well as the rate of change of CTE with respect to temperature for the Al–Si alloy, was significantly lower than that of unreinforced aluminium alloys (~23 μm/(m·°C)), thus providing a relatively excellent CTE match to copper. The prior heat-treatment condition of this Al–Si alloy was seen to have negligible influence over the temperature range tested.

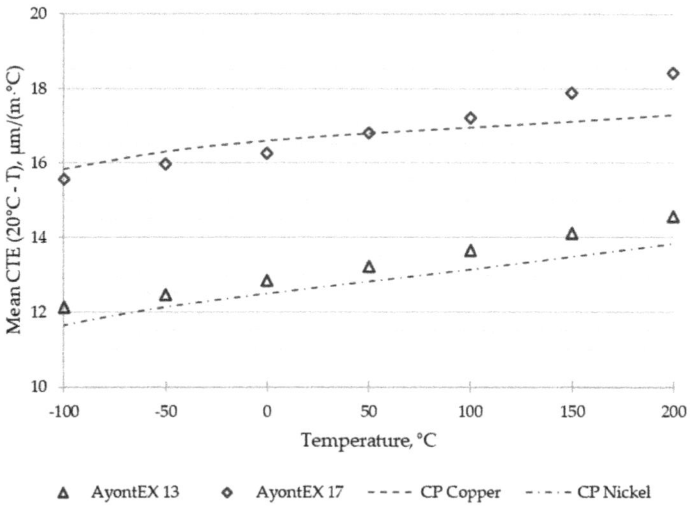

Figure 6. Achieved CTE matches to reference data on commercially pure copper and nickel.

A close CTE match was observed to nickel, with the 42% Al–Si alloy displaying a marginally higher CTE for the temperatures tested. The rate of change of CTE with respect to temperature for this Al–Si alloy and nickel was similar over the temperature range tested. Again, no influence of the prior heat treatment condition was observed.

3.3. *Additional Physical and Mechanical Properties*

Beyond CTE, the additional key material properties of the tested Al–Si alloys are presented in Table 1. A good balance of strength and conductivity for thermal management applications was displayed by these alloys. Given the large ranges in strength for AyontEX 17 quoted in Table 1, typical engineering stress–strain curves for this alloy are provided ii Figure 7 and true stress–strain equivalents can be seen in Figure 8.

Table 1. Summary of achieved typical properties. Ranges indicate dependence on heat treatment.

Property		Unit	AyontEX 17	AyontEX 13
Composition		-	6063 + 28% Si	6063 + 42% Si
Density		g/cm^3	2.60	2.54
Elastic Modulus		GPa	87	103
Specific Stiffness		Gpa/g/cm^3	33	41
Mean CTE	(−100–20 °C)	μm/(m·°C)	15.6	12.1
	(20–50 °C)		16.8	13.2
	(20–100 °C)		17.2	13.7
	(20–200 °C)		18.4	14.6
Thermal Conductivity		W/mK	160–170	134
Specific Heat Capacity		J/g/K	0.88	0.85
0.2% Proof Strength		Mpa	170–300	300–340
Ultimate Tensile Strength		Mpa	240–355	325–345
Specific Strength		Mpa/g/cm^3	92–137	128–136

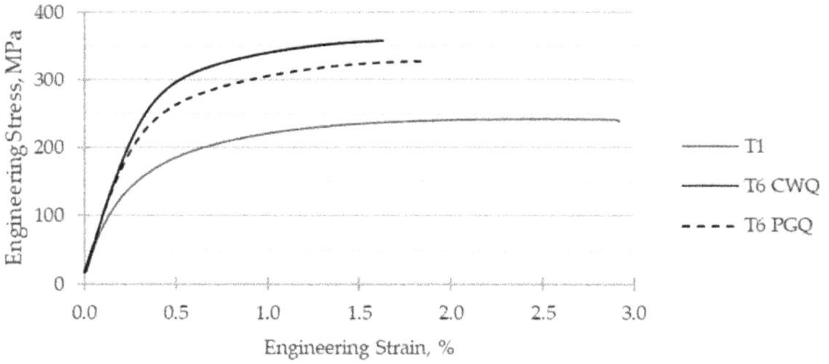

Figure 7. Typical room temperature engineering tensile stress–strain curves for AyontEX 17 alloy in different heat treatment conditions. T1 = air cooled from HIP, CWQ = cold water quench, PGQ = 25% polyglycol solution quench.

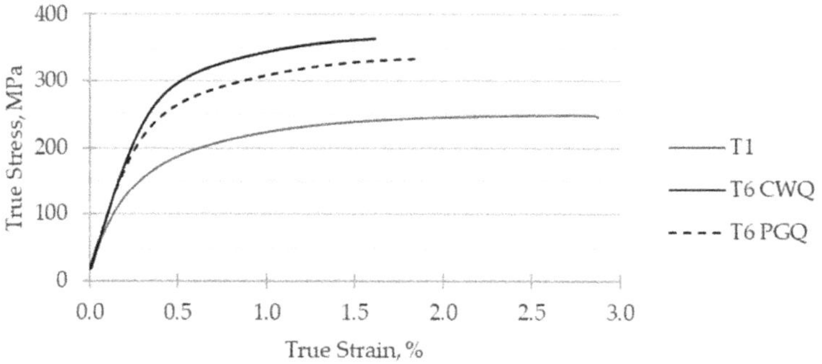

Figure 8. Typical room temperature tensile true stress–strain curves for AyontEX 17 alloy in different heat treatment conditions. T1 = air cooled from HIP, CWQ = cold water quench, PGQ = 25% polyglycol solution quench.

4. Discussion

The simple rule of mixtures calculation model provided a close approximation to the measured relationship between CTE and Si content, as shown by Figure 3. The measured values were consistently slightly lower than predicted and trended away from the model at lower and higher Si content values. A possible explanation for this deviation between the model and the measured values is that the rule of mixtures calculation does not account for phases other than the constituent elements (e.g., Mg_2Si precipitates), or the different elastic properties of the matrix, reinforcement, and associated thermal strains [40]. Particularly at the more extreme values of Si reinforcement ($\geq$50%), the so-called percolation threshold may have been passed, meaning it would be inappropriate to consider the Si phase and the elastic region surrounding it to be individual particles, but more likely a continuous path of reinforcement [41]. This could explain the increased deviation from the model observed at the highest Si contents tested. The negligible effect of heat treatment condition on the CTE of these alloys can be explained by the high Si content necessary to achieve such low CTEs. This thereby limits the ability to influence properties via heat treatment.

The compositions of the two now-commercially-available alloys, AyontEX 17 (with a CTE of 17 μm/(m·°C)) at 28% Si and AyontEX 13 (with a CTE of 13 μm/(m·°C)) at 42% Si, were derived from the measured CTE curve in Figure 3 as described in Section 3. The fine and homogeneous microstructures achieved (Figures 4 and 5) are important in enabling the application of high-throughput and available downstream machining and finishing processes to these alloys. Ease of machinability relative to equivalent liquid-state processed Al–Si alloys or other lightweight, low-CTE materials (e.g., MMCs) can provide a significant benefit in overall part cost by increasing throughput and decreasing tooling costs. Such benefits can be clearly realized in the manufacture of complex, high-performance components, where fine surface finish and high tolerance is required. Several common machining processes, including milling, turning, drilling, tapping, and electrical discharge machining (EDM), have been successfully performed on this material, many of which are highlighted by Figure 9. Here, the thin wall sections at the top of the part are 0.8 mm thick, and the EDM-processed slot at bottom is 0.5 mm thick and 32 mm deep. These features were achieved with standard, high-throughput methods using no special parameters.

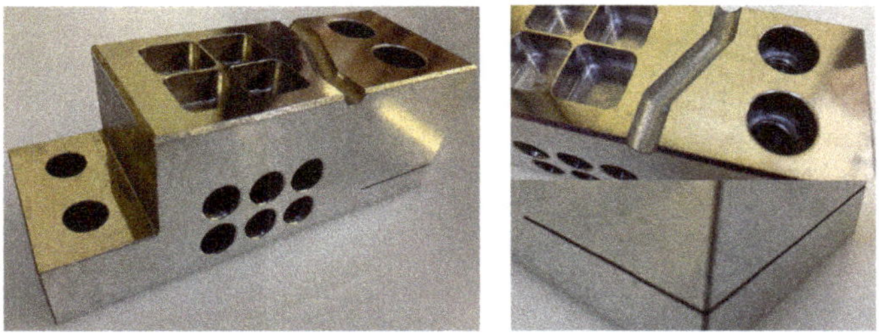

Figure 9. Demonstration of machining process on AyontEX 17 material, highlighting ability to machine thin-wall structures and tap holes (**top-right**) and EDM-processed 0.5 mm slot (**left**).

Example components representing typical applications for each of the alloys are presented by Figure 10. The miniature structural heatsink in Figure 10a is 58 mm × 40 mm × 5 mm in size, highlighting the ability to carry out precise milling, drilling, and tapping operations with the AyontEX 17 material. The AyontEX 13 mirror substrate presented by Figure 10b was approx. 150 mm in diameter. This was produced via rough machining using standard carbide tooling and completed with poly-crystalline diamond tooling. Given the low-density (2.54 g/cm^3) and complex lightweighting geometry, this mirror had a mass of 0.2 kg. Such lightweight mirror design is highly beneficial for airborne applications. More detailed demonstration of lightweight mirror manufacture in this material and associated application-specific testing has been covered elsewhere [42,43].

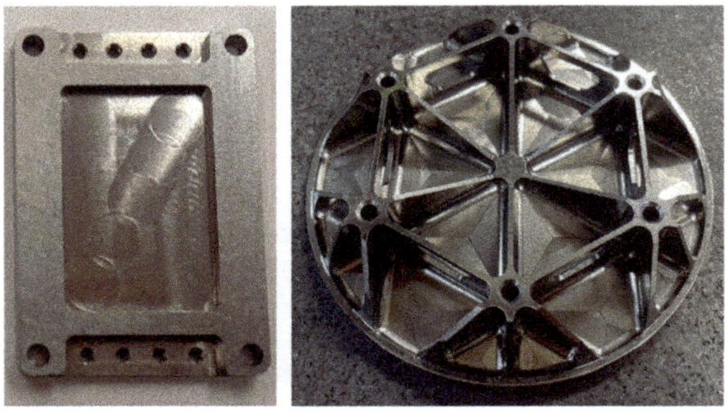

(**a**) (**b**)

Figure 10. Example applications of the Al–Si alloys. (**a**) Structural heat sink manufactured in AyontEX 17. (**b**) Optical mirror substrate manufactured in AyontEX 13.

An emerging topic in lightweight optical mirror design is additive manufacturing to enable complex structures for material and load optimization [44–46]. It is then an interesting topic for future exploration to develop and apply additive manufacturing techniques to the AyontEX 13 material, given this alloy already exists in a powder form, as per the powder metallurgy route presented here. Such development would combine the material property benefits discussed here with the design freedom of additive manufacturing for high-performance applications.

Close CTE matches with mating materials are important in the applications discussed above. The results presented in Figure 6 confirm that precise CTE matches to the reference data for copper and nickel materials were achieved with the manufactured alloys over the full −100–200 °C test range. Particularly, for AyontEX 13, the rate of change of CTE with respect to temperature was very consistent with that of commercially pure nickel. This is important for reflective optical and instrument systems, where precise CTE matches to nickel plating layers are critical for both dimensional and thermal stability over broad operating conditions [2–5,42,43]. It should be noted that in the case of electroless plating, the nickel material will contain some level of phosphorus. The CTE curve for NiP [3,4,47], is very similar to the commercially pure nickel CTE data used for comparison in Figure 6. It can therefore be said the AyontEX 13 material provides an excellent CTE match to both electrolytic and electroless Ni plating for application in high-precision optical mirrors. Of course, the ability to Ni plate the material is critical for such applications. An example of successful application of electrolytic nickel plating to AyontEX 13 is given by Figure 11. Electroless NiP plating solutions are also compatible [43], should this be preferred for the application.

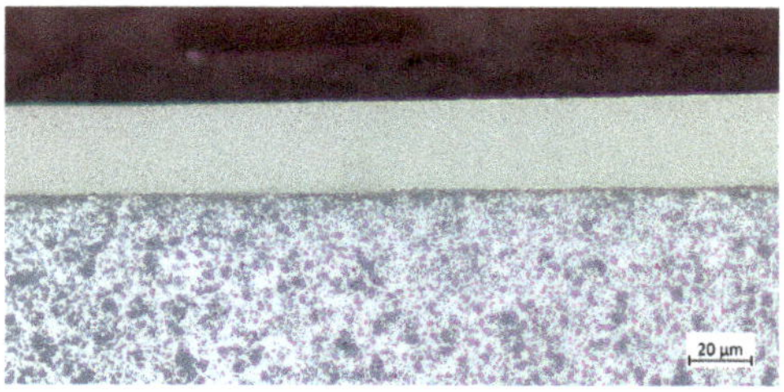

Figure 11. Optical micrograph of AyontEX 13 with electrolytic nickel plating applied, at 500× magnification.

The prior heat treatment condition of the material (T1 v. T6) was seen to have negligible impact on CTE for all compositions tested. It can therefore be understood that controlling the thermal expansion of these hypereutectic Al–Si alloys is best achieved through adjusting the Si content, as opposed to via heat treatment optimization.

As expected, increasing the Si content resulted in a linear decrease in density and a linear increase in elastic modulus. Increasing Si content was seen to result in a decrease in thermal conductivity, but also a decrease in the dependence of heat treatment condition on the thermal conductivity. For AyontEX 17 (6063 + 28% Si), the mean thermal conductivity was determined to be 160 W/mK in the T6 condition, but 170 W/mK in the T1 (air-cooled from HIP) condition, a difference of ~6%. In contrast, for AyontEX 13 (6063 + 42% Si), the heat treatment condition was found to have a negligible impact on the thermal conductivity.

This high thermal conductivity of AyontEX 17 in the T1 condition, coupled with the precise CTE match to Cu over the 0–100 °C temperature range (see Figure 6), highlights the materials' applicability for use in lightweight structural heatsink assemblies. As a direct replacement for 6061 Al, for example, AyontEX 17 provides an increased modulus and decreased CTE and density, whilst maintaining thermal conductivity. This is critical in enabling high power density devices, where heat load must be efficiently managed, as well as interfacial stresses driven by CTE mismatches to mating Cu components.

Similar trends with respect to Si loading were observed regarding the mechanical properties, and in particular the yield strength. The 0.2% proof strength of AyontEX 17 was seen to almost double from 170 MPa in T1 to 300 MPa in T6, with significant differences in ultimate tensile strength and strain to failure. These differences are depicted by the stress–strain curves given by Figures 7 and 8. This data was generated on billet material directly following the HIP process. The application of secondary forming processes, such as forging and extrusion, will enhance the ductility of the finished product. This remains an interesting topic for further exploration to allow close to shape forming operations. By comparison, the mechanical properties in the higher Si vol% materials had a much lower dependence on the heat treatment condition, as outlined by Figures 12 and 13. These effects can be explained by simply considering that as the Si vol% was increased, the vol% of the 6063-alloy decreased. This in turn meant that less Mg was available within the alloy for Mg_2Si precipitation strengthening effects through heat treatment. Additionally, as the quantity of the primary Si phase particles was increased, the relative distance between adjacent Si particles was decreased, and thus strengthening because of the primary Si phase became a more dominant mechanism relative to the Mg_2Si precipitation phases.

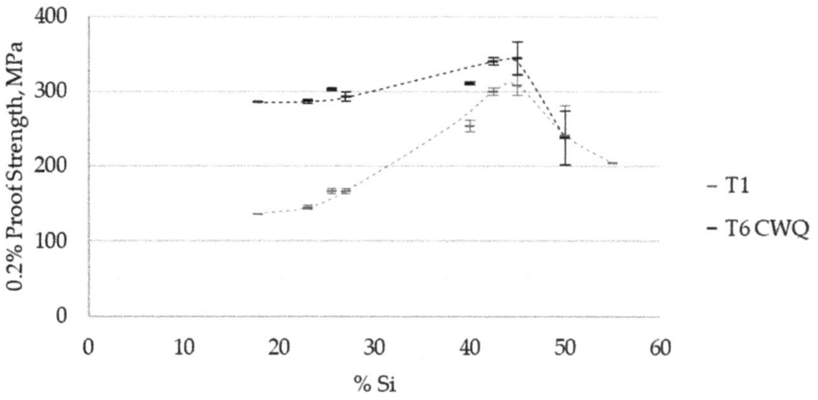

Figure 12. Relationship between 0.2% proof strength versus Si vol% and heat treatment condition. Data points are presented as mean values with 90% confidence intervals, where multiple tests were completed. T1 = air cooled from HIP, CWQ = cold water quench.

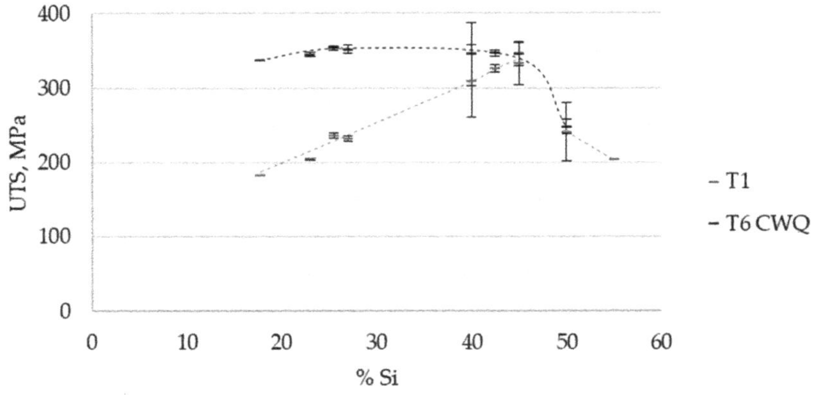

Figure 13. Relationship between ultimate tensile strength versus Si vol% and heat treatment condition. Data points are presented as mean values with 90% confidence intervals, where multiple tests were completed. (T1 = air cooled from HIP, CWQ = cold water quench).

The increasing influence of the Si phase as the Si vol% was increased to more extreme values ($\geq$50%) led the ensuing material to become more brittle. This resulted in an associated decrease in both 0.2% proof strength and ultimate tensile strength, as captured by Figures 12 and 13. For such compositions, the proportion of the primary Si phase is so significant that fracture can occur at the alloy–Si particle interface, due to Si particle clustering or the percolation effects described above. The severity of such effects within particulate MMC materials is well understood [48]. In contrast, at a more modest Si vol%, such as those used for the AyontEX 17 and AyontEX 13 materials, fracture occurs predominantly through the ductile 6063 alloy matrix phase.

Deeper exploration of the microstructural effects and phases at play in these hypereutectic Al–Si alloys provides an interesting avenue for future study. Investigating the impacts of using recycled Al alloy as an input raw material, and an associated high Fe content, could prove interesting. Previous studies have identified effective Si to have a significant impact on the performance of 6000-series aluminium alloys with high Fe content [49]. Exploration of how this effect translates to the more extreme Si levels in these hypereutectic alloys would be an interesting topic, particularly so with a view towards possible incorporation of recycled or secondary aluminium sources in the AyontEX alloys or SupremEX MMCs to increase the circularity of these high-performance materials.

5. Conclusions

A family of now commercially available high-performance hypereutectic Al–Si alloys (AyontEX™) have been developed with precise CTE matches to copper and nickel. Core material properties for the intended applications have been characterized, along with the key manufacturability considerations. The fine and homogeneous microstructure provides enhanced performance and manufacturability relative to equivalent liquid-state processed alloys. Further development work for these Al–Si alloys should be directed towards the use of mechanical alloying technology for lower CTE values, as well as testing the CTE of existing alloys over a broader temperature range. The development of additional fabricated forms via forging and extrusion for increased ductility and product form flexibility is also under investigation. Exploration of the incorporation of secondary or recycled aluminium sources as a raw material in this manufacturing process, and subsequent understanding of the effects of associated impurity elements, is an area for possible future study also.

Another potential application for Materion's mechanical alloying technology and these alloys moving forward is additive manufacturing. Particularly for lightweight, high-precision optical mirror manufacture, the AyontEX 13 composition is interesting. More broadly, however, the mechanical alloying process employed in this study allows for compositions to be tailored to suit the requirements for additive methods in terms of alloy and reinforcement composition and microstructures. The process also has sufficient scalability to provide an industrial solution. Due to this ability to mix a wide range of materials to produce light metal composite powders, it may be possible to produce new and novel powders as an input for additive processes. This could lead to additive manufacturing material and property options that do not exist today.

These research areas provide exciting opportunities for further exploitations of mechanical alloying technology, and powder metallurgy in general, towards developing new products to meet the needs of customers in global markets such as within the aerospace, high-performance automotive, space, and defense sectors.

Author Contributions: Conceptualization, A.T., A.F., F.G., N.F. and M.A.; Methodology, P.L., A.T. and F.G.; Project Administration, P.L. and A.F.; Investigation, P.L. and J.N.; Formal Analysis, P.L. and A.T.; Visualization, P.L.; Validation, F.G.; Writing—Original Draft Preparation, P.L.; Writing—Review & Editing, P.L., A.T., A.F., F.G., N.F. and M.A.; Supervision, A.T. and A.F. All authors have read and agreed to the published version of the manuscript.

Funding: This research received no external funding.

Data Availability Statement: The original contributions presented in the study are included in the article, further inquiries can be directed to the corresponding author.

Acknowledgments: The authors thankfully acknowledge the support provided by Outpost Technologies Inc. in the manufacture of the example mirror component presented in Figure 10b.

Conflicts of Interest: Authors P.L., A.T. and J.N. were employed by company Materion UK Limited, Author A.F. was employed by Materion Brush GmbH and Authors F.G., N.F. and M.A. were employed by Materion Corporation. We declare that the research was conducted in the absence of any commercial or financial relationships that could be construed as a potential conflict of interest.

References

1. Acreman, M.; Lewis; Farrah, N. Advanced Materials for Defense Applications-Part 2: Thermal Management. 2022. Available online: https://www.materion.com/en/insights/blog/advanced-materials-for-defense-applications-part-2-thermal-management (accessed on 7 May 2024).
2. Hibbard, D.L. Electroless Nickel for Optical Applications. *Adv. Mater. Opt. Precis. Struct. A Crit. Rev.* **1997**, *10289*, 173–199.
3. Rohloff, R.-R.; Gebhardt, A.; SchÃjnherr, V.; Risse, S.; Kinast, J.; Scheiding, S.; Peschel, T. A novel athermal approach for high-performance cryogenic metal optics. In Proceedings of the SPIE 7739, Modern Technologies in Space- and Ground-Based Telescopes and Instrumentation, San Diego, CA, USA, 20 July 2010.
4. Kinast, J.; Grabowski, K.; Gebhardt, A.; Rohloff, R.-R.; Risse, S.; Tünnermann, A. Dimensional Stability of Metal Optics on Nickel Plate AlSi40. In Proceedings of the SPIE 10563 International Conference on Space Optics, Tenerife, Spain, 6–10 October 2014.
5. Kinast, J.; Tünnermann, A.; Undisz, A. Dimensional Stability of Mirror Substrates Made of Silicon Particle Reinforced Aluminum. *Materials* **2022**, *15*, 2998. [CrossRef] [PubMed]
6. Zhang, R.; Zou, C.; Wei, Z.; Wang, H. Effect of High Pressure and Temperature on the Evolution of Si Phase and Eutectic Spacing in Al-20Si Alloys. *Crystals* **2021**, *11*, 705. [CrossRef]
7. Mostafa, A.; Alshabatat, N. Microstructural, Mechanical and Wear Properties of Al–1.3%Si Alloy as Compared to Hypo/Hyper–Eutectic Compositions in Al–Si Alloy System. *Crystals* **2022**, *12*, 719. [CrossRef]
8. Jiandon, P.; Talangkun, S. Microstructural Modification Hardness and Surface Roughness of Hypereutectic Al–Si Alloys by a Combination of Bismuth and Phosphorus. *Crystals* **2022**, *12*, 1026. [CrossRef]
9. Rosso, M. Ceramic and Metal Matrix Composites: Route and Properties. *J. Mater. Process. Technol.* **2006**, *175*, 364–375. [CrossRef]
10. Ujah, C.O.; Kallon, D.V.V. Trends in Aluminium Matrix Composite Development. *Crystals* **2022**, *12*, 1357. [CrossRef]
11. Kruzhanov, V.; Arnhold, V. Energy Consumption in Powder Metallurgical Manufacturing. *Powder Metall.* **2012**, *55*, 14–21. [CrossRef]
12. Vine, W.J.; Goodwin, P.S. Feasibility of Synthesising Lightweight Nanophase Al Materials by Mechanical Alloying. *Mater. Sci. Forum* **2000**, *331–337*, 1145–1150. [CrossRef]
13. Benjamin, J.S. Mechanical Alloying. *Sci. Am.* **1976**, *234*, 40–48. [CrossRef]
14. Suryanarayana, C.; Al-Aqeeli, N. Mechanically alloyed nanocomposites. *Prog. Mater. Sci.* **2012**, *58*, 383–502. [CrossRef]
15. Narayan, S.; Rajeshkannan, A. Workability behavior of powder metallurgy carbide reinforced aluminum composites during hot forging. *Mater. Manuf. Process.* **2015**, *30*, 1196–1201. [CrossRef]

16. Zhou, H.; Zhang, C.; Han, B.; Qiu, J.; Qin, S.; Gao, K.; Liu, J.; Sun, S.; Zhang, H. Microstructures and Mechanical Properties of Nanocrystalline AZ31 Magnesium Alloy Powders with Submicron TiB2 Additions Prepared by Mechanical Milling. *Crystals* **2020**, *10*, 550. [CrossRef]
17. Umeda, J.; Nishimura, N.; Fujii, H.; Jia, L.; Kondoh, K. In-Situ Formed Al3Zr Compounds Reinforced Al Composites and Tribological Application. *Crystals* **2021**, *11*, 227. [CrossRef]
18. Wu, Y.; Luo, S.; Wu, J.; Guo, B.; Wu, Z.; Chen, B.; Yu, Z.; Zhang, Z.; Li, W. Development and Characterization of CrCoNi Medium Entropy Alloy Particles Reinforced Aluminum Matrix Composite. *Crystals* **2022**, *12*, 1452. [CrossRef]
19. Arora, G.S.; Saxena, K.K.; Mohammed, K.A.; Prakash, C.; Dixit, S. Manufacturing Techniques for Mg-Based Metal Matrix Composite with Different Reinforcements. *Crystals* **2022**, *12*, 945. [CrossRef]
20. Ariff, A.M.M.; Lin, O.J.; Jung, D.-W.; Tahir, S.M.; Sulaiman, M.H. Rice Husk Ash as Pore Former and Reinforcement on the Porosity, Microstructure, and Tensile Strength of Aluminum MMC Fabricated via the Powder Metallurgy Method. *Crystals* **2022**, *12*, 1100. [CrossRef]
21. Fan, M.; Zhao, F.; Liu, Y.; Yin, S.; Peng, S.; Zhang, Z. Zinc Matrix Composites Reinforced with Partially Unzipped Carbon Nanotubes as Biodegradable Implant Materials. *Crystals* **2022**, *12*, 1110. [CrossRef]
22. Kushwaha, A.K.; Misra, M.; Menezes, P.L. Effect of Magnesium Dopant on the Grain Boundary Stability of Nanocrystalline Aluminum Powders during Cryomilling. *Crystals* **2023**, *13*, 541. [CrossRef]
23. Sübütay, H.; Şavklıyıldız, İ. Effect of High-Energy Ball Milling in Ternary Material System of (Mg-Sn-Na). *Crystals* **2023**, *13*, 1230. [CrossRef]
24. Yan, Q.; Chen, B.; Zhou, X.; Kondoh, K.; Li, J. Effect of Metal Powder Characteristics on Structural Defects of Graphene Nanosheets in Metal Composite Powders Dispersed by Ball Milling. *Crystals* **2021**, *11*, 260. [CrossRef]
25. Nava-Dino, C.G.; Flores-De los Ríos, J.P.; Maldonado-Orozco, M.C.; Sánchez-Carrillo, M.; Bautista-Margulis, R.G.; Delgado AD, L.C.; Almeraya-Calderón, F. Electrochemical Noise Response of Cr2Nb Powders Applying Mechanical Alloying. *Crystals* **2022**, *12*, 482. [CrossRef]
26. Gharsallah, H.I.; Azabou, M.; Khitouni, M.; Daza, J.; Suñol, J.-J. Study of the Microstructural, Thermal, and Magnetic Properties of High-Energy Ball-Milled Nanocrystalline Fe(Al). *Crystals* **2022**, *12*, 1430. [CrossRef]
27. ASM International. Milling of Brittle and Ductile Materials. In *ASM Handbook, Volume 7 Powder Metallurgy*; ASM International: Detroit, MI, USA, 2015; pp. 77–87.
28. Trautmann, M.; Ahmad, H.; Wagner, G. Influencing the Size and Shape of High-Energy Ball Milled Particle Reinforced Aluminum Alloy Powder. *Materials* **2022**, *15*, 3022. [CrossRef]
29. Raducanu, D.; Cojocaru, V.D.; Nocivin, A.; Hendea, R.E.; Ivanescu, S.; Stanciu, D.; Trisca-Rusu, C.; Serban, N.; Drob, S.I.; Campian, R.S. Microstructure Evolution during Mechanical Alloying of a Biodegradable Magnesium Alloy. *Crystals* **2022**, *12*, 1641. [CrossRef]
30. Hashiguchi, D.; Tricker, D.; Tarrant, A. Mechanically Alloyed Aluminum Metal Matrix Composites. In Proceedings of the SPIE 10372 Material Technologies and Applications to Optics Structures Components and Sub-Systems III, San Diego, CA, USA, 5 September 2017.
31. Frehn, A.; Lewis; Tarrant, A. Partikelverstärkte Aluminium-Werkstoffe für Hochleistungsanwendungen. In Proceedings of the Tagungsband 40. Hagener Symposium, Pulvermetallurgie—Vielfältige Prozesse und Werkstoffe, Hagen, Germany, 25 November 2022; pp. 309–336.
32. Hashiguchi, D.; Tricker, D.; Tarrant, A.; Campbell, J.; Pokross, C. Discontinuously reinforced aluminum MMC extrusions. *Met. Powder Rep.* **2017**, *72*, 252–258. [CrossRef]
33. Reiff-Musgrove, R.; Gaiser-Porter, M.; Gu, W.; Campbell, J.; Lewis; Frehn, A.; Tarrant, A.; Tang, Y.; Burley, M.; Clyne, T.W. Indentation Plastometry of Particulate Metal Matrix Composites, Highlighting Effects of Microstructural Scale. *Adv. Eng. Mater.* **2023**, *25*, 2201479. [CrossRef]
34. Winter, L.; Hockauf, K.; Lampke, T. Temperature and Particle Size Influence on the High Cycle Fatigue Behavior of the SiC Reinforced 2124 Aluminum Alloy. *Metals* **2018**, *8*, 43. [CrossRef]
35. ASTM International. Standard Test Method for Thermal Diffusivity by the Flash Method, E1461-13. In *ASTM Volume 14.01: Quality and Statistics; Hazard Potential of Chemicals; Thermal Measurements; Manufacture of Pharmaceutical and Biopharmaceutical Products; Healthcare Informatics*; ASTM International: West Conshohocken, PA, USA, 2022.
36. Nix, F.C.; MacNair, D. The Thermal Expansion of Pure Metals: Copper, Gold, Aluminum, Nickel and Iron. *Phys. Rev.* **1941**, *60*, 597–605. [CrossRef]
37. Hidnert, P. Thermal Expansion of Copper and Some of its Important Industrial Alloys. *Sci. Pap. Bur. Stand.* **1921**, *17*, 91–159. [CrossRef]
38. Hidnert, P. Thermal Expansion of Some Nickel Alloys. *J. Res. Natl. Bur. Stand.* **1957**, *58*, 89–92. [CrossRef]
39. Bennett, S.J. The thermal expansion of copper between 300 and 700 K. *J. Phys. D Appl. Phys.* **1978**, *11*, 777–780. [CrossRef]
40. Morrell, R. *Thermal Properties of Composite Materials-Measurements, Models and Thermal Exposure Derived Changes in MMCs and CMCs*; NPL Report CMMT (A) 6; National Physical Laboratory Management Ltd.: Teddington, UK, 1995.
41. Sergo, V.; Meriani, S. Thermal Expansion and Percolation in a SiC Whisker-Reinforced Ceramic Composite. *Mater. Sci. Lett.* **1991**, *10*, 855–857. [CrossRef]

42. Few, A.; Echols, J. Aluminium metal matrix composites: A capable low-cost mirror substrate. In Proceedings of the SPIE 12677 Astronomical Optics: Design, Manufacture, and Test of Space and Ground Systems IV, San Diego, CA, USA, 4 October 2023.
43. Sweeney, M.; Colling, S.; Schmidt, J.; Few, A.; Echols, J.; Reardon, P.; Douthit, D.; Farrah, N.; Lewis, P.; Tarrant, A.; et al. Application and Testing of SupremEX®640XA and AyontEX™ 13 for Mirrors and Precision Structures, In Proceedings of the SPIE 13042 Advanced Optics for Imaging Applications: UV through LWIR IX, National Harbor, MD, USA, 23 April 2024.
44. Zhang, K.; Hemeng, Q.; Guan, H.; Zhang, J.; Zhang, X.; Xiaolin, X.; Yan, L.; Wang, C. Design and Fabrication Technology of Metal Mirrors Based on Additive Manufacturing: A Review. *Appl. Sci.* **2021**, *11*, 10630. [CrossRef]
45. Zhang, K.; Xie, X.; Wang, C.; Wang, H.; Xu, F.; Wang, H.; Zhang, X.; Guan, H.; Qu, H.; Zhang, J. Optomechanical Performances of Advanced Lightweight Mirrors Based on Additive Manufacturing. *Mircomachines* **2022**, *13*, 1334. [CrossRef] [PubMed]
46. Zhang, J.; Wang, C.; Qu, H.; Guan, H.; Wang, H.; Zhang, X.; Xie, X.; Wang, H.; Zhang, K.; Li, L. Design and Fabrication of an Additively Manufactured Aluminum Mirror with Compound Surfaces. *Materials* **2022**, *15*, 7050. [CrossRef] [PubMed]
47. Fujimori, Y.; Shimizu, M.; Kurashina, T.; Arai, S. Substrate Thermal Expansion Coefficient Effect on Cracks Induced by the High-heat Treatment of Electroplated Ni-P Films for Power Devices. *Mater. Lett.* **2023**, *350*, 134869. [CrossRef]
48. Murphy, A.; Howard, S.; Clyne, T. Characterisation of severity of particle clustering and its effect on fracture of particulate MMCs. *Mater. Sci. Technol.* **1998**, *14*, 959. [CrossRef]
49. Österreicher, J.; Arnoldt, A.; Gneiger, S.; Kunschert, G. Tolerance of Al–Mg–Si Wrought Alloys for High Fe Contents: The Role of Effective Si. *Metall. Mater. Trans. A* **2023**, *54A*, 4472–4480. [CrossRef]

crystals

MDPI

Article

Microstructural Evolution and Thermal Stability of Long Period Stacking Ordered Phases in $Mg_{97}Er_2Ni_1$ and $Mg_{97}Er_2Zn_1$ Alloys

Jian Yin [1,2], Yushun Liu [2,*] and Guo-Zhen Zhu [2,*]

[1] Key Laboratory of Jiangxi Province for Modern Agricultural Equipment, School of Mechanical and Electrical Engineering, Jinggangshan University, Ji'an 343009, China; yinjian@jgsu.edu.cn

[2] Department of Mechanical Engineering, Manitoba Institute for Materials, University of Manitoba, Winnipeg, MB R3T 5V6, Canada

* Correspondence: liuy11@myumanitoba.ca (Y.L.); guozhen.zhu@umanitoba.ca (G.-Z.Z.)

Abstract: The influence of transition metals (Ni and Zn) on the formation, morphology, and thermal stability of long-period stacking ordered (LPSO) phases in $Mg_{97}Er_2Ni_1$ and $Mg_{97}Er_2Zn_1$ alloys was investigated. In the as-cast state, both alloys consist of α-Mg and LPSO phases. Heat treatment at 540 °C for 20 h dissolves block-like and lamellar LPSO phases into the α-Mg matrix in the $Mg_{97}Er_2Zn_1$ alloy, with lamellar LPSO phases reprecipitating during subsequent cooling from 540 °C to 400 °C. Comparative analysis shows that Ni significantly enhances the thermal stability of the LPSO phase compared to Zn. Ni favors the formation of block-shaped LPSO phases, while Zn facilitates lamellar LPSO precipitation within the α-Mg matrix. The LPSO phase in the $Mg_{97}Er_2Ni_1$ alloy exhibits an exceptionally high melting temperature of 605 °C, the highest reported for an LPSO phase. Additionally, heat treatment at 500 °C for 100 h preserves the area fraction of the LPSO phase in the $Mg_{97}Er_2Ni_1$ alloy, and at 540 °C for 100 h, the LPSO grains grow without phase dissolution or structural transformation of their 18R-type configuration. These findings provide valuable insights into the role of alloying transition metal elements in controlling the stability and morphology of LPSO phases, offering pathways for tailoring the morphology of the LPSO phase in the Mg-based alloys.

Keywords: magnesium; Mg-Er-Ni; Mg-Er-Zn; long period stacking order phase (LPSO)

1. Introduction

Magnesium (Mg) alloys are widely valued for their lightweight nature and excellent specific strength, making them ideal for applications in aerospace, automotive, and biomedical fields [1–5]. Among Mg-based alloy systems, Mg-RE (rare earth metal) alloys are renowned for their high strength and excellent corrosion resistance [3,4]. The addition of transition metals (TMs, such as Zn, Ni, or Cu) [6–15] or aluminum (Al) [16–18] to Mg-RE alloys can result in the formation of a secondary phase with a long-period stacking ordered (LPSO) structure. The LPSO phase imparts Mg-RE-TM or Mg-RE-Al alloys with unique mechanical properties [19–23] and corrosion behaviors [15,24–30]. As secondary phases, the LPSO phases influence the mechanical and corrosion properties of Mg-RE-TM or Mg-RE-Al alloys through their structure, morphology, amount, and distribution.

The morphology of the LPSO phase, typically manifesting as block-like or lamellar shapes, is a critical factor influencing the mechanical and corrosion properties of Mg-based alloys. Mechanically, block-like LPSO phases distributed along grain boundaries promote isometric crystal grains, enhancing ductility, as observed in Mg-Gd-Er-Zn-Zr alloys, while lamellar LPSO phases within grain interiors form strip-shaped grains, improving strength [19]. In $Mg_{94}Y_3Ni_3$ (at%) alloys, block-shaped LPSO phases aligned along the extrusion direction support the alloys to achieve a tensile yield strength of up to 502 MPa [20]. Similarly, in Mg-Y-Zn alloys, the blocky LPSO phase strengthens the alloy through strong basal fiber texture and deformation kink bands [21]. In contrast, the lamellar LPSO phase enhances ultimate tensile strength and elongation via kink deformation while simultaneously improving damping capacity and mechanical properties in Mg-Gd-Y-Zn-Zr-Nd alloys [22]. Regarding corrosion behavior, the continuous distribution of blocky LPSO phases with moderate volume fractions minimizes corrosion rates in Mg-Zn-Y alloys [25], and lamellar LPSO phases promote uniform corrosion, reducing rates in Mg-Gd-Zn-Zr alloys [26]. However, lamellar LPSO phases can increase corrosion rates in Mg-Gd-Ni alloys by raising the cathodic phase fraction [27]. Morphological changes in blocky phases, such as transitioning from a continuous network to disordered rods in Mg-7.2Y-2.8Ni alloys, raise degradation rates by 27% [28]. Similarly, in Mg-Er-Gd-Ni alloys, continuous blocky networks act as corrosion barriers, while streamlined LPSO phases weaken this effect [29]. In MgGd3Ni3 (at%) alloys, transforming continuous blocky LPSO phases into discontinuous rod-like and lamellar forms increases corrosion rates by 58.1%, driven by changes in corrosion pathways and increased galvanic corrosion sites [30]. These findings highlight the key role of LPSO morphology in optimizing the mechanical and corrosion performance of Mg-based alloys.

Heat treatment is an important strategy to tailor the formation and morphology of the LSPO phases [19,22,31–36]. Ding [34] demonstrated that in the $Mg_{96.32}Gd_{2.5}Zn_1Zr_{0.18}$ alloy, a heat treatment at 500 °C transformed $(Mg,Zn)_3Gd$ into blocky LPSO phases. Wang [19] reported a similar transformation of the $(Mg,Zn)_3(Gd,Er)$ phase at grain boundaries into blocky LPSO phases in as-cast Mg-Gd-Er-Zn-Zr alloys after heat treatment between 400 °C and 515 °C. Li [35] demonstrated that in the Mg-10Gd-3Y-1.2Zn-0.4Zr (wt%) alloy, heat treatment at 525 °C gradually dissolved blocky LPSO phases. Dang [22] showed that in the Mg-10Gd-2Y-1Zn-0.5Zr-0.2Nd (wt%) alloy, no LPSO phase formed after heat treatment at 540 °C for 4 h, but lamellar LPSO phases emerged during subsequent heat treatment at 450 °C for 10 h. Wu [36] observed that in a Mg-10Gd-1Zn-0.5Zr (wt%) alloy, solution treatment at 460 °C for 12 h eliminated the lamellar LPSO phase, forming blocky LPSO phases at grain boundaries. Further heat treatment at 500 °C for 12 h dissolved the blocky phases, and subsequent solution treatment at 420 °C for 1 h resulted in uniform lamellar LPSO phases within the α-Mg matrix. These findings also imply that thermal stability of the LPSO phase at elevated temperatures is crucial in determining the morphology of the LPSO phase during the heat treatment. Yin [10] obtained a highly stable blocky LPSO phase, which can hardly dissolve into the α-Mg matrix at 500 °C, by introducing Ni element into the Mg-Gd alloy, suggesting that the thermal stability of the LPSO phase strongly depends on Ni element. Liu [8] reported that a replacement of Ni for Zn in the $Mg_{94}Y_4Zn_2$ (at%) alloy resulted in an increase in the melting temperature of the LPSO phase from 547.5 °C to 571.1 °C, showing thermal stability enhancement for the LPSO phase. However, the underlying mechanism of how Ni or Zn affects the formation and thermal stability of the LPSO phase remains unclear.

Recently, Mg-Er-Ni alloys have garnered significant interest due to their high strength and rapid corrosion rate, making them promising candidates for fracture tools in the oil industry [37–40]. The mechanical properties and corrosion behaviors of Mg-Er-Ni alloys are strongly influenced by the morphology of the LPSO phase. Therefore, Mg-Er-Ni alloy was chosen as the material of study to compare with Mg-Er-Zn alloy, aiming to better understand how Ni and Zn influence the morphology of the LPSO phase during heat treatment, offering new insights into tailoring LPSO phase morphology.

2. Experimental Procedure

The $Mg_{97}Er_2Ni_1$ and $Mg_{97}Er_2Zn_1$ (at%) alloys were fabricated using pure magnesium (Mg), erbium (Er), nickel (Ni), and zinc (Zn), each with a purity exceeding 99.9 wt%. According to the weight ratio of the alloy composition, pure metal materials of Mg, Er, Ni, or Zn were prepared. After removing surface oxides from the pure metal materials, the metals were placed into a graphite crucible. Using high-frequency induction heating, the metals were melted and thoroughly mixed. The temperature was maintained at approximately 750 °C for 30 min, followed by natural cooling to room temperature in the graphite crucible to obtain the as-cast alloy ingot. The samples, with dimensions of approximately 10 mm × 10 mm × 8 mm, were cut from the as-cast alloy ingot for heat treatment. The heat treatments were conducted in an argon-protected furnace.

X-ray diffraction (XRD; D8 Advance, Bruker AXS, Karlsruhe, Germany) was performed using monochromatic Cu Kα radiation for phase identification at a scanning speed of 10°/min. Scanning electron microscopy (SEM; FEI Quanta 200, Hillsboro, OR, USA), equipped with energy-dispersive spectroscopy (EDS; Oxford IE250X-Max50, Abingdon, Oxfordshire, UK), was used to analyze the microstructure and phase compositions. Transmission electron microscopy (TEM; FEI Tecnai G2 F20, Eindhoven, The Netherlands) was employed to identify the structure of the LPSO phase. Differential thermal analysis (DTA; Diamond TG/DTA, PerkinElmer, Shelton, CT, USA) was conducted at a heating rate of 10 °C/min to assess the thermal behavior of the alloys.

SEM samples were polished using SiC grinding paper and then etched for less than 10 s in a 4% nitric acid in ethanol solution before observation. Energy-dispersive spectroscopy (EDS) analysis was conducted using an acceleration voltage of 20 kV and a working distance of about 10 mm. TEM samples were prepared by wire-cutting the ingot into thin slices (~1 mm thick), mechanically grinding them to a thickness of ~100 μm, and then punching out 3 mm diameter disks from the pre-ground slices, which were then thinned using an ion beam for final preparation. The area fraction of the LPSO phase was evaluated by counting the number of pixels in the SEM images.

3. Results and Discussion

3.1. Dual-Phase Microstructure in As-Cast Alloys

Both as-cast $Mg_{97}Er_2Ni_1$ and $Mg_{97}Er_2Zn_1$ alloys exhibit the dual-phase microstructure, as confirmed by X-ray diffraction (XRD) and electron microscopy (EM) characterization. Figure 1 shows the XRD patterns of as-cast $Mg_{97}Er_2Ni_1$ and $Mg_{97}Er_2Zn_1$ alloys, with the peak presence corresponding to both α-Mg and LPSO phase. The corresponding SEM images and EDS results are shown in Figure 2, presenting dendrite-like α-Mg matrix (dark area) and interconnected networks of block-like LPSO phase (light area). The measured area fractions of block-like LPSO phases in the network were ~39% for the $Mg_{97}Er_2Ni_1$ alloy and ~18% for the $Mg_{97}Er_2Zn_1$ alloy.

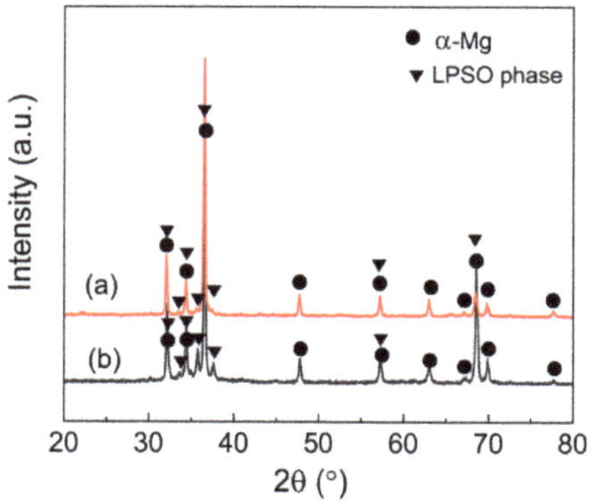

Figure 1. XRD patterns of as-cast (**a**) $Mg_{97}Er_2Ni_1$ and (**b**) $Mg_{97}Er_2Zn_1$ alloys.

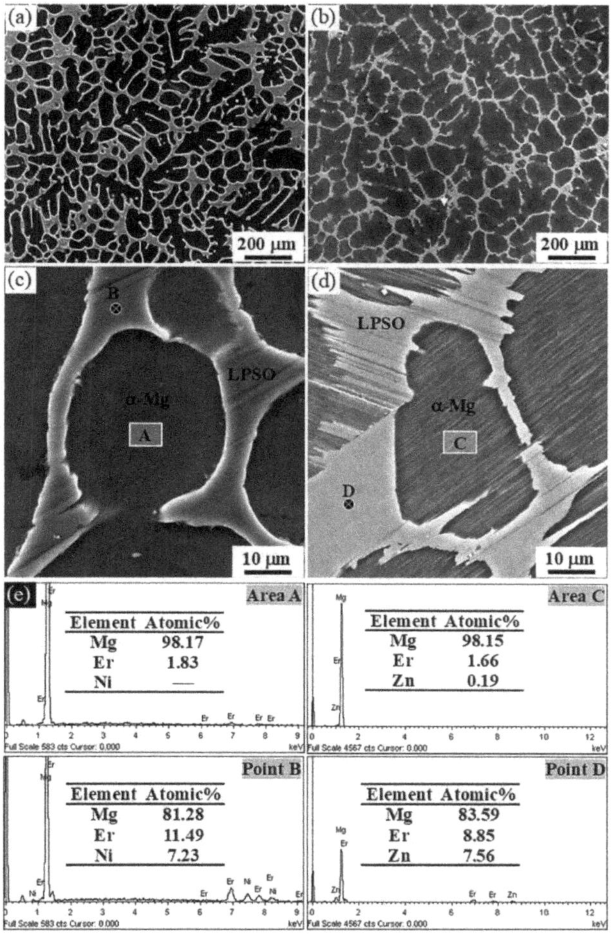

Figure 2. SEM images of the as-cast alloys: (**a**,**c**) $Mg_{97}Er_2Ni_1$ and (**b**,**d**) $Mg_{97}Er_2Zn_1$, along with (**e**) the corresponding EDS results.

Alongside interconnected networks of block-like LPSO phases, fine lamellar structures were observed within the α-Mg grains of the $Mg_{97}Er_2Zn_1$ alloy. These fine lamellar structures appear as needle-like features in light gray in Figure 2d, further supported by the corresponding EDX spectra shown in Figure 2e. The characteristics of lamellar LPSO phases embedded within the α-Mg grains suggest that the lamellar LPSO phase precipitates within the α-Mg grains during solidification. On the other hand, these lamellar structures were absent in the $Mg_{97}Er_2Ni_1$ alloy, as indicated by the morphology of Area A in Figure 2c and corresponding EDS results in Figure 2e, showing that no Ni element was detected in the α-Mg matrix.

It is known that the formation of LPSO structure requests the co-segregation of rare-earth (RE) elements and transition metals (TM, such as Ni or Zn) [6–15]. According to the Mg-Ni binary phase diagram, the maximum solid solubility of Ni in Mg is nearly zero across the temperature range from room temperature to 506 °C. In contrast, the Mg-Zn binary phase diagram shows a maximum solid solubility of Zn in Mg of 2.4 at% at 325 °C. As a result, during solidification, Ni atoms can be hardly incorporated into the α-Mg solid. This is consistent with the abovementioned fact that no Ni element was detected in the α-Mg matrix of the $Mg_{97}Er_2Ni_1$ alloy. However, Zn atoms could diffuse into α-Mg and enable the formation of the LPSO phase within the α-Mg grains. This is evident by the morphology of the lamellar LPSO phase within the α-Mg grains of the $Mg_{97}Er_2Zn_1$ alloy shown in Figure 2d and the EDS result of the α-Mg grain (Mg-1.66 at%Er-0.19 at%Zn) in Figure 2e. This highlights a critical influence of the solubility of transitional metal elements in α-Mg on tailoring LPSO morphology, that is, Ni promotes block-like LPSO phase formation, while Zn enables both block-like and lamellae-like LPSO phases.

3.2. Thermal Stability of Involved LPSO Phases

The thermal behaviors of as-cast $Mg_{97}Er_2Ni_1$ and $Mg_{97}Er_2Zn_1$ alloys were investigated using differential thermal analysis (DTA), as shown in Figure 3. Both alloys exhibit two endothermic peaks. According to our previous study [10], the first peak corresponds to the melting of the LPSO phase, and the second one means the melting of the α-Mg phase. It is noted that the melting temperature of the LPSO phase in the $Mg_{97}Er_2Ni_1$ alloy is as high as 605 °C, which is 57 °C higher than that of the LPSO phase in the $Mg_{97}Er_2Zn_1$ alloy. This represents the highest melting temperature for an LPSO phase reported to date, significantly exceeding 528 °C for the LPSO phase in a Mg-Gd-Zn alloy [34], 550 °C for the LPSO phase in Mg-Y-Zn alloys [41], and 540 °C for the LPSO phase in Mg-Gd-Ni alloys [10].

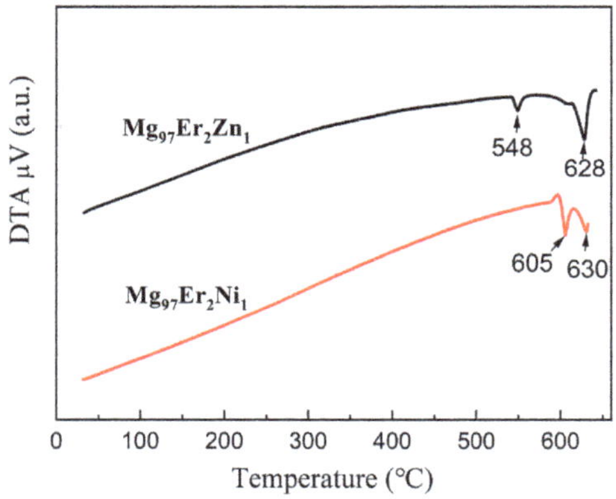

Figure 3. DTA curves of the as-cast $Mg_{97}Er_2Ni_1$ and $Mg_{97}Er_2Zn_1$ alloys.

The second peak, corresponding to the melting of α-Mg, is almost identical for both as-cast $Mg_{97}Er_2Ni_1$ and as-cast $Mg_{97}Er_2Zn_1$ alloys, suggesting that the RE and TM solutes have negligible influence on the melting of α-Mg. As the thermal analysis was conducted on the as-cast alloys, the phase transitions observed during heating can be considered the reverse process of solidification. During solidification, Mg enriched with RE and TM elements to its solubility limit initially precipitates as the dendritic α-Mg matrix from the molten liquid. It should be noted that Ni, which has very limited solubility in α-Mg, tends to stay exclusively in the remaining liquid, in contrast to Zn, which distributes between the α-Mg and the remaining liquid. The remaining liquid solidifies to form the LPSO phase between the α-Mg dendrites, resulting in a blocky LPSO network. In the meantime, within the α-Mg grains of as-cast $Mg_{97}Er_2Zn_1$ alloy, the lamellar LPSO phase precipitates as the presence of both Er and Zn. The thermal stability of LPSO phases and the speculated phase transitions were further examined by characterizing the microstructure and its evolution over the heat treatment temperatures ranging from 500 °C to 540 °C.

3.3. Microstructure Evolution During Heat Treatment

To investigate the microstructure evolution of $Mg_{97}Er_2Ni_1$ and $Mg_{97}Er_2Zn_1$ alloys during heat treatment, the as-cast alloys were subjected to heat treatments at temperatures near the phase transition ranges of typical LPSO phases. The two alloys were heat-treated at 540 °C for 20 h and then cooled down to 400 °C in the furnace, followed by quenching in the water.

Figure 4 presents the SEM images and corresponding EDS results of the heat-treated $Mg_{97}Er_2Ni_1$ and $Mg_{97}Er_2Zn_1$ alloys at 540 °C for 20 h. Compared to the as-cast alloy, the heat-treated $Mg_{97}Er_2Ni_1$ alloy is still composed of two phases of α-Mg and LPSO phase. The area fraction of the block-like LPSO phase is about ~37%, almost the same as the as-cast alloy; however, the morphology of the LPSO slightly changes, with the edges becoming more rounded. No Ni element was detected within the α-Mg matrix, as evidenced by the EDS result from Area A in Figure 4e. In contrast, the morphology of the LPSO phase has significant changes in the heat-treated $Mg_{97}Er_2Zn_1$ alloy, where the as-cast bulky LPSO phase network was broken. Enlarged SEM image (Figure 4d) and EDS result (Figure 4e) show that except for α-Mg and LPSO phases, there are Zn-rich particle phase formed, as illustrated by the black arrows in Figure 4d. Interestingly, no Zn was detected in α-Mg matrix, as evidenced by the composition of Mg-2.67 at%Er within Area E in α-Mg of the heat-treated $Mg_{97}Er_2Zn_1$ alloy. This suggests that during the heat treatment at 540 °C, Zn atoms tend to segregate into the LPSO phase or Zn-rich particle phase instead of α-Mg. Compared to the as-cast $Mg_{97}Er_2Zn_1$ alloy, no lamellar structure was observed in the α-Mg, as shown in Figure 4d, which should be attributed to the absence of Zn in the α-Mg matrix. This indicates that the heat treatment at 540 °C could enable the lamellar the LPSO phase to dissolve into the α-Mg matrix. In addition, the area fraction of the block-like LPSO phase in the heat-treated $Mg_{97}Er_2Zn_1$ alloy was measured to be ~9%, which is much less than ~18% for the as-cast $Mg_{97}Er_2Zn_1$ alloy. This suggests that the block-like LPSO phase in the $Mg_{97}Er_2Zn_1$ alloy is unstable and could dissolve into α-Mg.

Figure 5 shows the SEM images and corresponding EDS results of the $Mg_{97}Er_2Ni_1$ and $Mg_{97}Er_2Zn_1$ alloys cooled in the furnace to 400 °C after a heat treatment at 540 °C for 20 h. No obvious change was observed in the $Mg_{97}Er_2Ni_1$ alloy. However, in the $Mg_{97}Er_2Zn_1$ alloy, within the α-Mg grains, lamellar structure was observed, which suggests that the lamellar LPSO phase precipitates during the cooling from 540 °C to 400 °C. It is expected that upon cooling from 540 °C, the Zn atoms in the LPSO phase will diffuse into the α-Mg, enabling the formation of lamellar LPSO phase within α-Mg grains. As demonstrated above, a proper heat treatment could regulate not only the morphology of the LPSO phase but also its spatial distribution in the $Mg_{97}Er_2Zn_1$ alloy.

3.4. High Thermal Stability of LPSO Phase in a $Mg_{97}Er_2Ni_1$ Alloy

As indicated in the DTA analysis in Figure 2, the LPSO phase in the $Mg_{97}Er_2Ni_1$ alloy exhibits high thermal stability, corresponding to its phase transition temperature. To further confirm this, as-cast $Mg_{97}Er_2Ni_1$ samples were heat-treated at 500 °C and 540 °C for 100 h each.

Figure 6 compares the microstructures of the as-cast $Mg_{97}Er_2Ni_1$ alloy (denoted as the ACNi alloy), the alloy after heat treatment at 500 °C for 100 h (denoted as the 500Ni alloy), and the alloy after heat treatment at 540 °C for 100 h (denoted as the 540Ni alloy). The microstructure of the 500Ni alloy exhibits slight differences compared to the ACNi alloy, with both alloys showing nearly identical area fractions of the LPSO network (~39%). However, a detailed inspection of the enlarged SEM images in Figure 6b,d reveals that the edges of the LPSO phases in the 500Ni alloy are more rounded, suggesting that a degree of homogenization has occurred during the heat treatment at 500 °C.

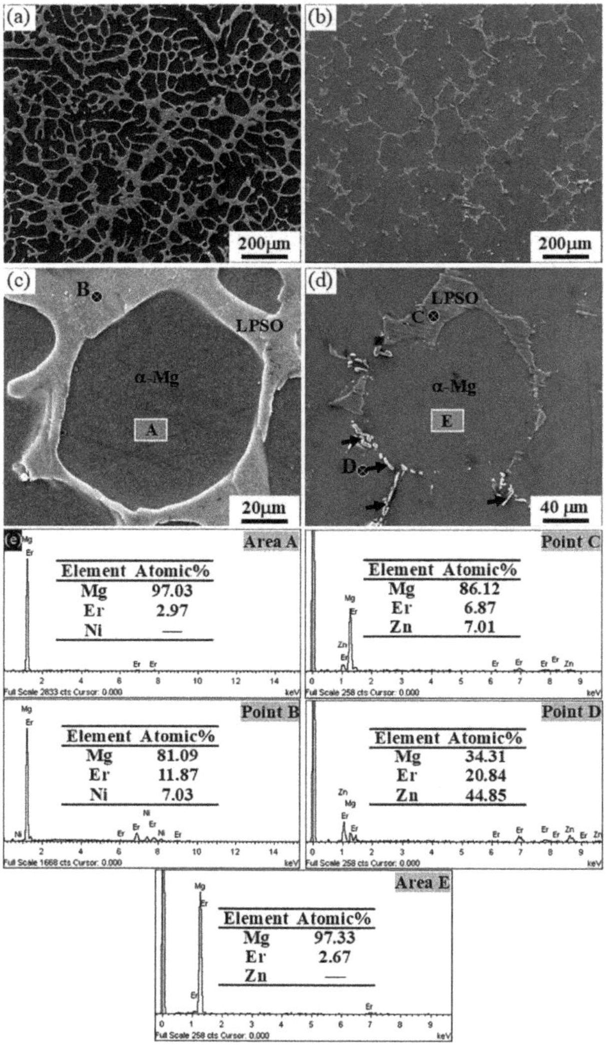

Figure 4. SEM images of (**a**,**c**) $Mg_{97}Er_2Ni_1$ and (**b**,**d**) $Mg_{97}Er_2Zn_1$ alloys after heat treatment at 540 °C for 20 h, along with (**e**) the corresponding EDS results.

In contrast, the microstructure of the 540Ni alloy shows significant changes compared to the ACNi and 500Ni alloys. Specifically, the number of thin LPSO network branches decreases markedly, and these thin branches are rarely observed in the 540Ni alloy. This indicates that the thinner LPSO structures are less stable at 540 °C. These observations suggest that the LPSO phase in the $Mg_{97}Er_2Ni_1$ alloy is relatively stable at 500 °C but undergoes noticeable change at 540 °C.

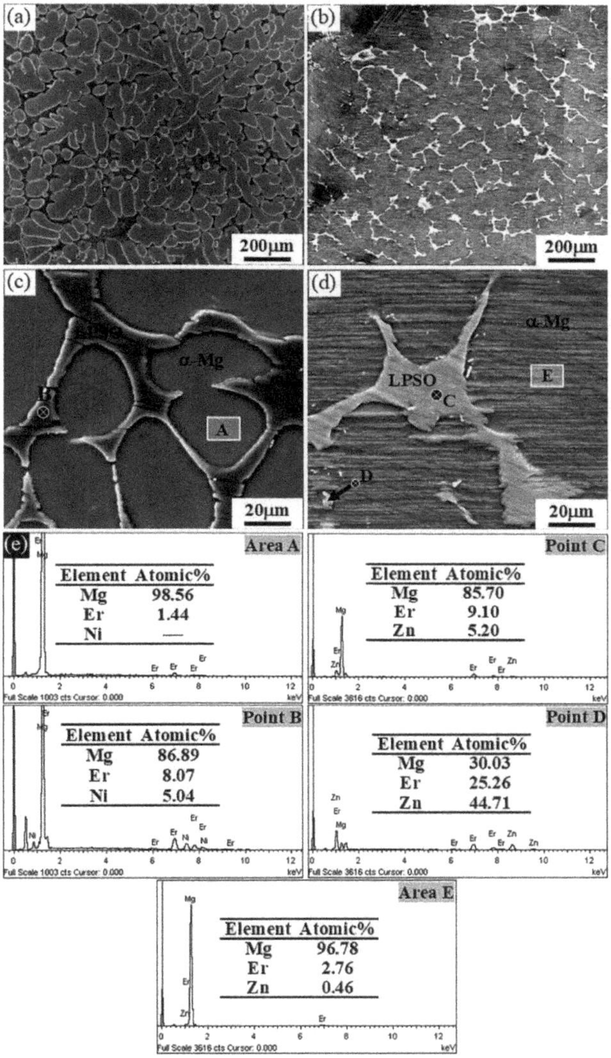

Figure 5. SEM images of (**a**,**c**) $Mg_{97}Er_2Ni_1$ and (**b**,**d**) $Mg_{97}Er_2Zn_1$ alloys after heat treatment at 540 °C for 20 h and subsequent furnace cooling to 400 °C followed by quenching in the water, along with (**e**) the corresponding EDS results.

The area fraction of the LPSO phase in the 540Ni alloy was measured to be ~36%, slightly lower than the ~39% observed for the ACNi and 500Ni alloys. This slight reduction in area fraction cannot be attributed to the dissolution of the LPSO phase into the α-Mg matrix. Figure 6g presents the EDS analysis of the 500Ni and 540Ni alloys, which indicates that no Ni was detected in the α-Mg matrix of either alloy, consistent with the results for the ACNi alloy shown in Figure 2e. This confirms that no Ni was dissolved into the α-Mg, and consequently, no LPSO phase dissolved into the matrix.

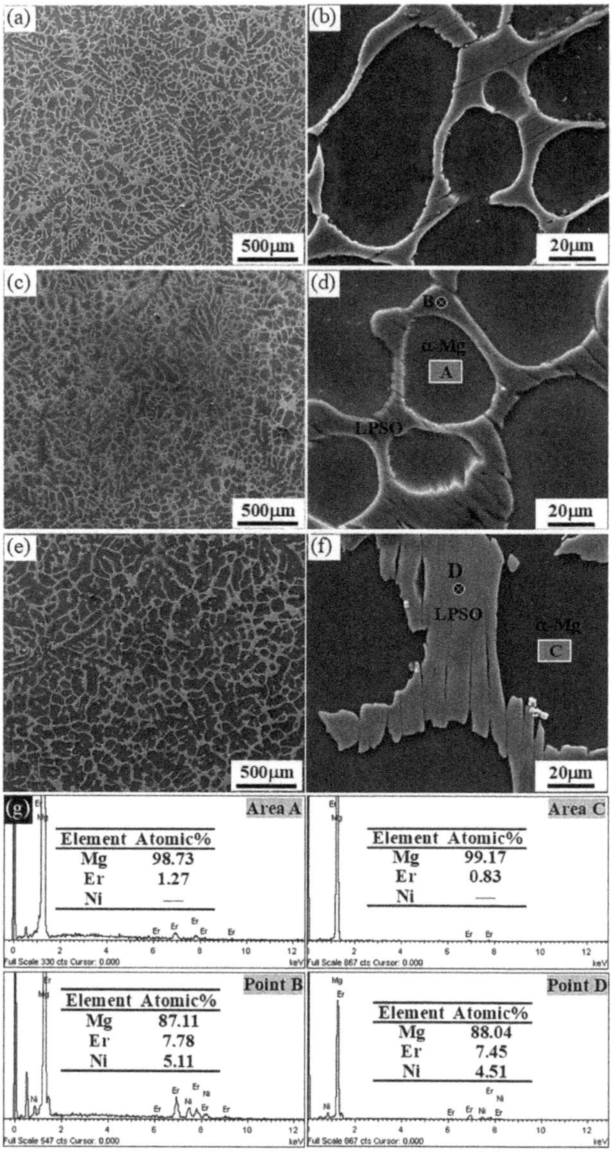

Figure 6. SEM images of $Mg_{97}Er_2Ni_1$ alloys: (**a**,**b**) in the as-cast condition, (**c**,**d**) after heat treatment at 500 °C for 100 h, and (**e**,**f**) after heat treatment at 540 °C for 100 h. The corresponding EDS results for the heat-treated alloys are presented in (**g**).

According to the Mg-Ni binary phase diagram, the eutectic temperature is 506 °C, and the solid solubility of Ni in Mg remains nearly zero from room temperature up to 506 °C. As a result, Ni atoms have very limited mobility and cannot easily diffuse into the α-Mg below 506 °C, which explains why the LPSO phase remains highly stable at 500 °C and why the thin LPSO branches persist even after 100 h of heat treatment.

Above 506 °C, however, Ni becomes more active and capable of diffusion. During heat treatment at 540 °C, this increased diffusion likely facilitates the growth of the LPSO phase. Thin LPSO branches may diffuse into adjacent LPSO structures, promoting their thickening and coalescence. This phenomenon is supported by the observation in Figure 6e, where the LPSO network branches in the 540Ni alloy appear wider than those in the ACNi and 500Ni alloys. Therefore, the slight reduction in the area fraction of the LPSO phase in the 540Ni alloy compared to the ACNi alloy can be attributed to short-range diffusion within the LPSO structure, leading to its growth and redistribution. This also suggested a new way to tailor the morphology of the LPSO phase by controlling the diffusion of TM solutes at varying temperatures.

The LPSO structures in the ACNi and 540Ni alloys were analyzed using TEM. Figure 7 presents high-resolution TEM images and the corresponding SAED patterns along the $[11\bar{2}0]_{Mg}$ direction of the LPSO phases. In both TEM images, the atomic arrangement exhibits a periodic interval of approximately 1.57 nm, corresponding to six close-packed atomic layers. These layers form one of three six-layer building blocks (ABABCA) that make up the unit cell of the 18R structure [7]. The SAED patterns for both alloys reveal five extra diffraction spots positioned between the direct beam and the $(0002)_{Mg}$ reflection, characteristic of the 18R structure [7]. These observations confirm that the LPSO phase in both the ACNi and 540Ni alloys maintains an 18R configuration. The consistency in the structure between the two alloys indicates that heat treatment at 540 °C does not alter the type of LPSO phase in the $Mg_{97}Er_2Ni_1$ alloy.

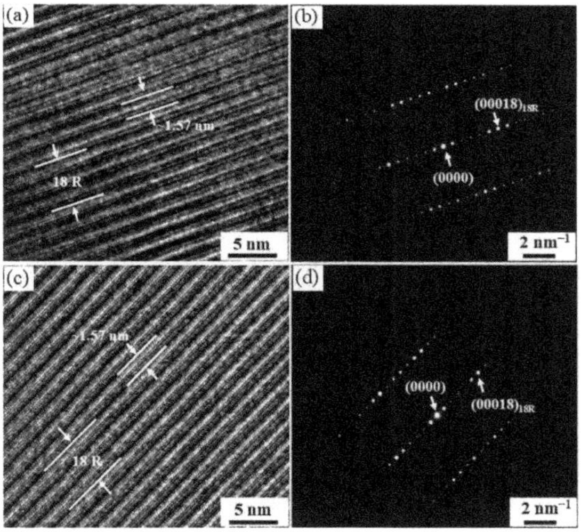

Figure 7. High-resolution TEM image and corresponding SAED pattern of the LPSO phase in the $Mg_{97}Er_2Ni_1$ alloys: (**a**,**b**) as-cast; (**c**,**d**) heat-treated alloys at 540 °C for 100 h. The electron beam is parallel to $[11\bar{2}0]$ of the α-Mg.

4. Conclusions

(1) In the as-cast state, both $Mg_{97}Er_2Ni_1$ and $Mg_{97}Er_2Zn_1$ alloys are composed of α-Mg and LPSO phases. Heat treatment at 540 °C for 20 h leads to the dissolution of both block-like and lamellar LPSO phases into the α-Mg matrix in the $Mg_{97}Er_2Zn_1$ alloy. During subsequent cooling from 540 °C to 400 °C, lamellar LPSO phases precipitate within the α-Mg grains.

(2) The LPSO phase in the $Mg_{97}Er_2Ni_1$ alloy exhibits an exceptionally high melting temperature of 605 °C, indicating superior thermal stability. After heat treatment at 500 °C for 100 h, the area fraction of the LPSO phase remains unchanged, underscoring its stability at 500 °C. At 540 °C, heat treatment promotes the growth of LPSO grains rather than dissolving the LPSO phase or altering its 18R-type structure. This indicates that elevated-temperature heat treatments can effectively modify the morphology of the LPSO phase without reducing its area fraction.

(3) Comparative heat treatment results demonstrate that Ni is more effective than Zn in enhancing the thermal stability of the LPSO phase. Ni promotes the formation of block-shaped LPSO phases, while Zn facilitates the precipitation of lamella-shaped LPSO phases within the α-Mg matrix, suggesting a way to tailor the morphology of the LPSO phase by adjusting the Ni or Zn elements.

These findings provide insights into the influence of alloying transition metal elements on the thermal stability and morphology of LPSO phases, offering potential pathways for tailoring the morphology of the LPSO phase in Mg-based alloys.

Author Contributions: The alloy samples were fabricated by J.Y.; the characterization was performed by J.Y. and Y.L.; data analysis was carried out by Y.L. and G.-Z.Z.; the manuscript was written by J.Y. and revised by Y.L. and G.-Z.Z. All authors have read and agreed to the published version of the manuscript.

Funding: J. Yin gratefully acknowledges the financial support of the National Natural Science Foundation of China (Grant No. 51661023) and the China Scholarship Council. Y. Liu and G.-Z. Zhu acknowledge the financial support of the NSERC Discovery Grant (RGPIN-2019-05882) and the Canada Research Chair program (CRC-2021-00512).

Data Availability Statement: The original contributions presented in this study are included in the article. Further inquiries can be directed to the corresponding author.

Conflicts of Interest: The authors declare no conflicts of interest.

References

1. Tan, J.; Ramakrishna, S. Applications of magnesium and its alloys: A review. *Appl. Sci.* **2021**, *11*, 6861. [CrossRef]
2. Kumar, L.G.S.; Thirumalaikumarasamy, D.; Karthikeyan, K.; Mathanbabu, M.; Ashokkumar, M.; Ramachandran, C.S. An overview of recent trends and challenges of post treatments on magnesium alloys. *Mater. Today Proc.* **2023**, *78*, 700–707. [CrossRef]
3. Wu, G.; Wang, C.; Sun, M.; Ding, W. Recent developments and applications on high-performance cast magnesium rare-earth alloys. *J. Magnes. Alloys* **2021**, *9*, 1–20. [CrossRef]
4. Zhang, J.; Liu, S.; Wu, R.; Hou, L.; Zhang, M. Recent developments in high-strength Mg-RE-based alloys: Focusing on Mg-Gd and Mg-Y systems. *J. Magnes. Alloys* **2018**, *6*, 277–291. [CrossRef]
5. Jayasathyakawin, S.; Ravichandran, M.; Baskar, N.; Chairman, C.A.; Balasundaram, R. Mechanical properties and applications of Magnesium alloy—Review. *Mater. Today Proc.* **2020**, *27*, 909–913. [CrossRef]
6. Geshani, M.S.; Kalayeh, P.M.; Asadi, A.H.; Mirzadeh, H.; Malekan, M.; Emamy, M. A review of Mg alloys containing long-period stacking ordered (LPSO) structures with insight into the application of friction stir processing. *J. Mater. Res. Technol.* **2023**, *24*, 4945–4966. [CrossRef]
7. Zhu, Y.M.; Morton, A.J.; Nie, J.F. The 18R and 14H long-period stacking ordered structures in Mg–Y–Zn alloys. *Acta Mater.* **2010**, *58*, 2936–2947. [CrossRef]
8. Liu, H.; Xue, F.; Bai, J.; Zhou, J.; Liu, X. Effect of substitution of 1 at% Ni for Zn on the microstructure and mechanical properties of $Mg_{94}Y_4Zn_2$ alloy. *Mater. Sci. Eng. A* **2013**, *585*, 387–395. [CrossRef]
9. Kawamura, Y.; Yamasaki, M. Formation and mechanical properties of $Mg_{97}Zn_1RE_2$ alloys with long-period stacking ordered structure. *Mater. Trans.* **2007**, *48*, 2986–2992. [CrossRef]
10. Yin, J.; Lu, C.; Ma, X.; Dai, B.; Chen, H.-L. Investigation of two-phase Mg-Gd-Ni alloys with highly stable long period stacking ordered phases. *Intermetallics* **2016**, *68*, 63–70. [CrossRef]
11. Liu, Y.; Yin, J.; Zhu, G. Solving Ambiguity in EBSD Indexing of Long-Period Stacking Ordered (LPSO) Phase in Mg with Template Matching Approach. *Crystals* **2024**, *14*, 932. [CrossRef]
12. Xu, H.; Jiang, J.; Bi, G.; Meng, S.; Zhang, X.; Li, Y.; Chen, T. Microstructure and mechanical properties of extruded and rolled Mg–Y–Zn–Ni–Co alloy. *J. Mater. Res. Technol.* **2023**, *27*, 8323–8333. [CrossRef]
13. Liu, H.; Huang, H.; Wang, C.; Sun, J.; Bai, J.; Xue, F.; Ma, A.; Chen, X.-B. Recent advances in LPSO-containing wrought magnesium alloys: Relationships between processing, microstructure, and mechanical properties. *JOM* **2019**, *71*, 3314–3327. [CrossRef]
14. Li, Y.; Yang, C.; Zeng, X.; Jin, P.; Qiu, D.; Ding, W. Microstructure evolution and mechanical properties of magnesium alloys containing long period stacking ordered phase. *Mater. Charact.* **2018**, *141*, 286–295. [CrossRef]
15. Xu, D.; Han, E.; Xu, Y. Effect of long-period stacking ordered phase on microstructure, mechanical property and corrosion resistance of Mg alloys: A review. *Prog. Nat. Sci. Mater. Int.* **2016**, *26*, 117–128. [CrossRef]
16. Zhu, Q.; Li, Y.; Zhang, H.; Wang, J.; Jiang, H.; Zhao, J. Revealing the Interaction between Dislocations and LPSO–Precipitates Structure in a Mg-Y-Al Alloy at Different Temperatures. *Crystals* **2024**, *14*, 1018. [CrossRef]
17. Kishida, K.; Yokobayashi, H.; Inui, H.; Yamasaki, M.; Kawamura, Y. The crystal structure of the LPSO phase of the 14H-type in the Mg–Al–Gd alloy system. *Intermetallics* **2012**, *31*, 55–64. [CrossRef]

18. Wang, Z.; Shen, Z.; Liu, Y.; Zhao, Y.; Zhu, Q.; Chen, Y.; Wang, J.; Li, Y.; Lozano-Perez, S.; Zeng, X. The effect of LPSO phase on the high-temperature oxidation of a stainless Mg-Y-Al alloy. *J. Magnes. Alloys* **2024**, *12*, 4045–4052. [CrossRef]
19. Wang, J.; Song, P.; Huang, S.; Pan, F. Effects of heat treatment on the morphology of long-period stacking ordered phase and the corresponding mechanical properties of Mg–9Gd–xEr–1.6 Zn–0.6 Zr magnesium alloys. *Mater. Sci. Eng. A* **2013**, *563*, 36–45. [CrossRef]
20. Wu, S.Z.; Qiao, X.G.; Qin, S.H.; Chi, Y.; Zheng, M. Improved strength in wrought Mg–Y–Ni alloys by adjusting the block-shaped LPSO phase and plate-shaped γ' phase. *Mater. Sci. Eng. A* **2022**, *831*, 142198. [CrossRef]
21. Yamasaki, M.; Hashimoto, K.; Hagihara, K.; Kawamura, Y. Effect of multimodal microstructure evolution on mechanical properties of Mg–Zn–Y extruded alloy. *Acta Mater.* **2011**, *59*, 3646–3658. [CrossRef]
22. Dang, C.; Wang, J.; Wang, J.; Yu, D.; Zheng, W.; Xu, C.; Lu, R. Effect of lamellar LPSO phase on mechanical properties and damping capacity in cast magnesium alloys. *J. Mater. Res. Technol.* **2023**, *22*, 2589–2599. [CrossRef]
23. Garces, G.; Barea, R.; Stark, A.; Schell, N. Anisotropic Plastic Behavior in an Extruded Long-Period Ordered Structure $Mg_{90}Y_{6.5}Ni_{3.5}$ (at.%) Alloy. *Crystals* **2020**, *10*, 279. [CrossRef]
24. Pałgan, D.; Dobkowska, A.; Zielińska, A.; Drozdenko, D.; Máthis, K.; Święszkowski, W. The Role of LPSO Structures in Corrosion Resistance of Mg-Y-Zn Alloys. *Crystals* **2022**, *12*, 1723. [CrossRef]
25. Zhang, J.; Xu, J.; Cheng, W.; Chen, C.; Kang, J. Corrosion behavior of Mg–Zn–Y alloy with long-period stacking ordered structures. *J. Mater. Sci. Technol.* **2012**, *28*, 1157–1162. [CrossRef]
26. Zhang, X.; Ba, Z.; Wang, Q.; Wu, Y.; Wang, Z.; Wang, Q. Uniform corrosion behavior of GZ51K alloy with long period stacking ordered structure for biomedical application. *Corros. Sci.* **2014**, *88*, 1–5. [CrossRef]
27. Ma, K.; Wang, J.; Peng, Y.; Dai, C.; Pan, Y.; Wang, D.; Wang, Y.; Pei, S.; Ma, Y. Enhanced degradation properties of Mg-Gd-Ni alloys by regulating LPSO morphology. *J. Phys. Chem. Solids* **2022**, *171*, 110974. [CrossRef]
28. Ma, K.; Wang, J.; Ren, J.; Dai, C.; Liu, S.; Peng, Y.; Pan, Y. Enhanced degradation properties of Mg-Y-Ni alloys by tailoring the LPSO morphology for fracturing tools applications. *Mater. Charact.* **2021**, *181*, 111489. [CrossRef]
29. Dai, C.; Zhang, S.; Wang, Y.; Wang, J.; Wang, J. Elucidation of the corrosion rate enhancement mechanism in Mg–Er–Gd–Ni alloys with high volume fraction of LPSO phase and different Gd contents after extrusion. *J. Mater. Res. Technol.* **2023**, *27*, 522–541. [CrossRef]
30. Dai, C.; Wang, J.; Pan, Y.; Ma, K.; Peng, Y.; Ren, J.; Wang, Y.; Wang, D.; Wang, J.; Ma, Y. Tailoring the microstructural characteristic and improving the corrosion rate of Mg-Gd-Ni alloy by heat treatment with different volume fraction of LPSO phase. *Corros. Sci.* **2023**, *210*, 110806. [CrossRef]
31. Li, Q.; Jin, L.; Wang, F.; Dong, S.; Zeng, J.; Wang, F.; Dong, J. Crack propagation path deflection induced by α-Mg/LPSO interface and its effect on the fracture toughness of cast Mg-RE-Zn alloys. *J. Magnes. Alloys* **2024**, *88*, 1–5. [CrossRef]
32. Zhang, G.; Niu, X.; Lu, R. Effects of heat treatment routes on microstructure and damping properties of Mg–Zn–Y–Mn alloys. *Mater. Res. Express* **2023**, *10*, 036515. [CrossRef]
33. Maier, P.; Schmahl, M.; Clausius, B.; Joy, C.; Fleck, C. Nanoindentation on the Transformation of LPSO Phases during Different Solution Heat Treatments in an Mg-Dy-Nd-Zn-Zr Alloy. *Crystals* **2022**, *12*, 1673. [CrossRef]
34. Ding, W.J.; Wu, Y.J.; Peng, L.M.; Zeng, X.; Yuan, G.; Lin, D. Formation of 14H-type long period stacking ordered structure in the as-cast and solid solution treated Mg-Gd-Zn-Zr alloys. *J. Mater. Res.* **2009**, *24*, 1842–1854. [CrossRef]
35. Li, D.; Zeng, X.; Dong, J.; Zhai, C. Influence of heat treatment on microstructure and mechanical properties of Mg-10Gd-3Y-1.2 Zn-0.4 Zr alloy. *Trans. Nonferrous Met. Soc. China* **2008**, *18*, s117–s121. [CrossRef]
36. Wu, X.; Pan, F.; Cheng, R.; Luo, S. Effect of morphology of long period stacking ordered phase on mechanical properties of Mg-10Gd-1Zn-0.5 Zr magnesium alloy. *Mater. Sci. Eng. A* **2018**, *726*, 64–68. [CrossRef]
37. Dai, C.; Wang, J.; Pan, Y.; Ma, K.; Peng, Y.; Wang, Y.; Wang, D.; Ran, C.; Wang, J.; Ma, Y. Achieving exceptionally high strength and rapid degradation rate of Mg-Er-Ni alloy by strengthening with lamellar γ' and bulk LPSO phases. *J. Mater. Sci. Technol.* **2024**, *168*, 88–102. [CrossRef]
38. Dai, C.; Pei, S.; Ma, K.; Wang, Y.; Wang, D.; Wang, J.; Ma, Y.; Wang, J. Rapid corrosion rates and high mechanical properties of as-extruded Mg–Er–Ni alloys by introducing LPSO and γ' phases. *J. Mater. Res. Technol.* **2023**, *24*, 6246–6263. [CrossRef]
39. Dai, C.; Pei, S.; Ma, K.; Wang, Y.; Wang, D.; Wang, J.; Ma, Y.; Wang, J. Revealing distinct corrosion mechanisms of soluble as-extruded Mg–Er–Ni alloy with LPSO and Mg_2Ni phase in different orientations. *J. Mater. Res. Technol.* **2023**, *26*, 1903–1921. [CrossRef]
40. Zhang, S.; Xie, X.; Ouyang, L.; Dai, C.; Li, H.; Xue, X.; Pei, S.; Wang, K.; Wang, J.; Pan, F. Thermo-mechanical processing characteristics and strain-rate-sensitive degradation properties of Mg-Er-Ni alloys. *J. Alloys Compd.* **2024**, *1004*, 175926. [CrossRef]
41. Chen, B.; Lin, D.; Zeng, X.; Lu, C. Effects of yttrium and zinc addition on the microstructure and mechanical properties of Mg–Y–Zn alloys. *J. Mater. Sci.* **2010**, *45*, 2510–2517. [CrossRef]

crystals

MDPI

Article

Multi-Scale Numerical Simulation of Short Tungsten Fiber Reinforced Tungsten–Copper Composites: Influence Mechanisms of Fiber Parameters

Longchao Zhuo [1,2,*], Yixing Xie [1], Hang Xu [2,3], Bin Luo [1], Nan Liu [1,4], Bingqing Chen [5,*] and Hao Wang [6,*]

[1] School of Materials Science and Engineering, Xi'an University of Technology, Xi'an 710048, China
[2] Department of Materials Science and Engineering, Kyoto University, Yoshida-honmachi, Sakyo-ku, Kyoto 606-8501, Japan
[3] College of Materials Science and Engineering, Hunan University, Changsha 410082, China
[4] School of Materials Science, Northwestern Polytechnical University, Xi'an 710072, China
[5] 3D Printing Research and Engineering Technology Center, AECC Beijing Institute of Aeronautical Materials, Beijing 100095, China
[6] Co-Creation Institute for Advanced Materials, Shimane University, 1060 Nishikawatsu, Matsue 690-8504, Japan
* Correspondence: zhuolongchao@xaut.edu.cn (L.Z.); hwtkjcbq1984@163.com (B.C.); hwang@mat.shimane-u.ac.jp (H.W.); Tel./Fax: +86-29-82312156 (L.Z.)

Abstract: Tungsten fiber reinforced tungsten–copper (W_f/W-Cu) composites have broad application prospects in fields such as electronic packaging due to their excellent comprehensive properties. However, the correlation between fiber parameters (content, aspect ratio, orientation) and the mechanical behavior of the materials is not yet clear. In this study, a combination of numerical simulation and experimental research was employed to construct a three-dimensional microstructural mechanic model and systematically investigate the influence of fiber parameters on the tensile properties and mechanisms of W_f/W-Cu composites. The results show that: (1) The critical fiber aspect ratio is 7.6. When below this value, fiber pullout dominates, and when above this value, fiber tensile fracture is the main mechanism. (2) As the fiber content increases from 1% to 6%, the tensile strength of the composite increases by 9.6%, the yield strength increases by 10.2%, while the elongation after fracture decreases by 18.6%. (3) As the fiber orientation angle increases from 0° to 90°, the material strength first increases and then decreases, while the toughness first decreases and then increases. (4) Short fibers achieve interface toughening through fiber pullout, crack deflection, and fiber bridging, while long fibers improve the strength and toughness of the composite through load transfer and fiber bridging effects. (5) The damage evolution mechanism reveals the regulation effect of fiber parameters on the multi-scale mechanical behavior of the material. The research results can guide the composition and structure optimization design of W_f/W-Cu composites, provide new ideas for the research of high-performance fiber composites, and have important significance for their engineering applications in extreme environments.

Keywords: W_f/W-Cu composites; fiber parameters; mechanical properties; microstructural mechanics simulation; damage evolution

1. Introduction

Metal matrix composites (MMCs) have been widely applied in various fields, such as aerospace, electronic information, energy conservation, and environmental protection, due to their excellent comprehensive properties [1–3]. As an important functional MMC, tungsten–copper (W-Cu) composites possess outstanding electrical and thermal conductivity and reliable high-temperature mechanical properties. They show broad application prospects in electrical contacts for high voltage switching, electronic packaging and thermal management, microwave and plasma applications, and aerospace and defense components, such as shaped charge liners and kinetic energy penetrators [4–7]. However, the mechanical properties of traditional W-Cu composites, especially the strength and plasticity under high-temperature conditions, cannot fully meet the requirements of harsh service environments.

As is known, the influence mechanisms of different reinforcement methods on the mechanical behavior of composites differ significantly. The mechanical properties of particle-reinforced composites mainly depend on the particle volume fraction and the particle/matrix interfacial bonding strength. However, particles have limited ability to bear large loads and have a limited strengthening and toughening effect on composites [8]. Fiber reinforcement is a technical approach that significantly improves the mechanical properties of composites by introducing high-strength, high-modulus fibers into the matrix. Compared with particle reinforcement, the unique geometric structure of fibers allows them to maximize their reinforcing effect along the load-bearing direction, greatly enhancing the specific strength and specific modulus of the material. Meanwhile, the fiber/matrix interface can achieve efficient load transfer between the two phases through stress transfer, further improving the damage tolerance of the composite [9]. Continuous fibers have a high aspect ratio and can fully exert their reinforcing effect in the fiber direction. However, continuous fibers are difficult to use in the preparation of complex-shaped components, and their mechanical properties exhibit high anisotropy, which to some extent limits their engineering applications [10,11]. Compared to continuous fibers and fiber mesh networks, short fibers offer advantages such as low-cost processing, flexible forming, and relatively weaker anisotropic mechanical properties, making them suitable for applications involving complex-shaped components and large-scale production. However, it is important to note that rationally designing the spatial distribution of short fibers is crucial for ensuring the optimal performance of short fiber reinforced composites in specific loading directions [12]. Numerous studies have shown that factors such as fiber content, aspect ratio, and orientation distribution have significant effects on the mechanical properties of composites, and the interfacial bonding strength between fibers and matrix is a key factor in determining material properties [13]. A. Kelly et al. [8] proposed the concept of critical fiber length based on the shear lag theory, pointing out that fibers can only effectively exert their reinforcing effect when their length exceeds the critical length. J.L. Thomason et al. [14] systematically investigated the influence of fiber content and aspect ratio on the mechanical properties of glass fiber/polypropylene composites and found an optimal matching range for fiber content and aspect ratio. These research works have provided important theoretical foundations and experimental evidence for the design and optimization of short fiber composites. In a recent review article on fiber-reinforced MMCs [15], various composite systems, such as SiC fiber-reinforced Ti-based and Al-based composites, stainless steel fiber-reinforced Al-based composites, Al_2O_3 fiber-reinforced NiAl-based composites, and Kevlar fiber-reinforced Mg-based composites, were summarized. Through single fiber pull-out experiments combined with finite element simulations, the fiber/matrix interfacial strength and its influencing factors were investigated. It was found that residual stress, interface thickness, shear strength, and fiber content have important effects on the interfacial properties and mechanical properties of composites, providing important references for the design and preparation of fiber-reinforced MMCs. However, it can be clearly seen that the above-mentioned work mainly focuses on light alloy matrix composites, while there are few reports on high-performance refractory metal matrix composites, such as W-Cu composites.

In 1973, Professor Kelly comprehensively summarized the mechanical behavior of short fiber reinforced metal matrix composites in his classic work 'Strong Solids' [16]. Based on the shear lag theory, he proposed Kelly's model, which provides universal theoretical guidance for the design and optimization of short fiber composites. However, Kelly's theory mainly relies on elastic mechanics analysis, and further refinement is needed to characterize mechanical behaviors such as plastic deformation and yield fracture. Moreover, the strengthening mechanisms and failure behaviors of different fiber reinforcement systems exhibit significant differences. Therefore, conducting systematic research on the strengthening and toughening mechanisms of short fibers in high-performance refractory metal matrices (such as tungsten–copper composites) is of great scientific significance and practical value for further developing multi-scale mechanical models and improving the quantitative characterization and prediction methods of mechanical properties of short fiber composites. In our previous study [17], we prepared short tungsten fiber (W_f) reinforced WCu composites and found that W_f with a preferential <110> orientation significantly enhanced the tensile strength of WCu composites, reaching a tensile strength of 549 MPa and plasticity of 10.2%. Preliminary three-dimensional finite element simulations revealed the influence of fiber fracture and interface debonding on material failure, but the effect of fiber parameters on the mechanical behavior of the material was not deeply analyzed. In summary, although some progress has been made in the field of short fiber-reinforced MMCs, and the effects of fiber parameters (content, aspect ratio) and interfacial bonding strength on the mechanical properties of materials have been preliminarily revealed, there is still a lack of systematic and in-depth research on short tungsten fiber-reinforced W-Cu composites, which have broad application prospects. Therefore, it is necessary to develop a multi-scale constitutive model based on physical mechanisms to quantitatively describe the influence of fiber parameters on the mechanical properties of materials and provide theoretical guidance for the structural design and performance optimization of short fiber-reinforced W-Cu composites.

In view of this, short tungsten fiber-reinforced W-Cu composites were selected as the research object in this study to systematically evaluate the influence of fiber content, aspect ratio, orientation, and other factors on the microstructure and macroscopic mechanical properties of the materials. Based on experimental research, we further constructed a three-dimensional numerical model containing the fiber/matrix interface. This model employs cohesive elements to describe the interfacial bonding performance and considers the nonlinear mechanical behavior of fibers and matrix, thus achieving a high-fidelity characterization of the microstructure and mechanical properties of short fiber composites. Through finite element numerical simulations, the effects of key parameters such as fiber content, aspect ratio, and orientation on the mechanical properties and failure behavior of the composites were quantitatively analyzed, revealing the multi-scale cooperative reinforcement mechanism of short fibers. Here, the selection of fiber volume fractions in composite materials requires a comprehensive consideration of various factors, including mechanical performance requirements, process feasibility, and cost. In general, a higher fiber volume fraction leads to increased specific strength and specific modulus of the composite. However, it also increases the tendency for fiber agglomeration and results in higher discreteness of the composite's mechanical properties [18]. Moreover, excessively high fiber content can significantly reduce the processability of the composite, causing issues such as insufficient matrix infiltration and increased residual stress [19]. Therefore, the selection of fiber volume fraction necessitates a balance between mechanical performance and processability. Considering the aforementioned research findings and the high-density characteristics of the tungsten–copper matrix used in this study, we selected a short tungsten fiber addition range of 1–6 wt.%. This content range not only fully exploits the strengthening and toughening effects of short fibers but also ensures good processability of the composite. Additionally, by analyzing the stress–strain distribution and damage evolution behavior during the tensile deformation of the composites, the failure mechanisms under different fiber parameters were clarified. The research results can provide an important theoretical basis for the structural design and performance optimization of short fiber-reinforced composites and have significant implications for expanding their engineering applications in fields such as electronic packaging.

2. Simulation Methods and Settings

In this study, the finite element method was employed to systematically investigate the mechanical and electrical properties and their influencing factors of short tungsten fiber-reinforced tungsten–copper composites. For the simulation of mechanical properties, Digimat-FE software [20] was used to generate the representative volume element (RVE) models [21] of the composites containing randomly oriented short fibers. Through comparative analysis, considering the complexity of the fiber network structure, tetrahedral elements were used to discretize the model. An average mesh size of 1 μm and an appropriate RVE size of 0.5 mm × 0.5 mm × 0.5 mm were selected to ensure that the model could accurately reflect the mechanical behavior of the composites. Subsequently, the RVE model was imported into ABAQUS for pre-processing operations such as finite element mesh generation, load and boundary condition application. Using this mechanical model, the effects of fiber content, aspect ratio, orientation angle, and other factors on the tensile properties of the composites were systematically simulated and analyzed. Specifically, this study focused on W_f/W-Cu composites with a fiber mass fraction of 4%. To accurately describe the microstructural characteristics of the composites, a three-dimensional geometric model containing randomly oriented fibers was established using Digimat-FE software and imported into ABAQUS for subsequent analysis. The W-Cu matrix was prepared using a random uniform distribution powder metallurgy process, so the size of the RVE model needed to be large enough to include the microstructural features of the material. Through Boolean operations, the fiber model was embedded into the matrix to obtain the fiber/matrix composite structure. Considering the important influence of the fiber/matrix interface bonding state on the properties of the composites, cohesive elements were used to describe the interfacial bonding properties. Unlike idealized rigid connections, cohesive elements can accurately simulate the damage initiation and evolution process of the interface. In the previous study [17], we obtained samples by ball-milling a mixture of short tungsten fibers with an average diameter of 10 μm and a length of 1 mm with W-15 wt.%Cu mixed powder and then using powder metallurgy techniques. The basic data used in the simulation comes from the experimental results. Please refer to Ref. [17] for detailed information. In terms of boundary conditions and loading, one end of the model was fixed, while a displacement load was applied to the other end until macroscopic fracture occurred. To avoid the influence of stress concentration on the results, a coupling constraint was applied to the surface where the displacement load was imposed to ensure uniform load distribution. Moreover, to accurately describe the elastoplastic deformation behavior of the materials, true stress–strain constitutive relationships were constructed for the fibers and matrix, respectively, based on tensile test data.

To accurately predict the mechanical behavior of the composites, an elastoplastic constitutive model considering damage evolution was adopted in this study. The elastic deformation stage of the material was controlled by the elastic modulus, and after yielding, it entered the plastic deformation stage. To simulate the damage and fracture behavior of the material, a ductile damage model was introduced. Figure 1a shows the stress–strain curve of the material considering damage evolution. The solid line in the figure represents the true stress–strain relationship considering damage effects, while the dotted line corresponds to the ideal elastoplastic curve without damage. σy0 and ε0pl are the yield stress and equivalent plastic strain corresponding to the moment of damage initiation, respectively. When the equivalent plastic strain increases ε0pl, the material enters the damage evolution stage, and a damage variable D ($0 \leq D \leq 1$) is introduced to characterize the degradation degree of material properties. D_{max} is taken as the critical value of 1. When $D = D_{max}$, the material element is considered to be completely failed and is deleted from the finite element model, thus simulating the initiation and propagation of cracks. The fiber/matrix interface is described using a bilinear cohesive model, as shown in Figure 1b. The cohesive elements are assumed to be elastic and do not undergo plastic deformation. The damage initiation criterion adopts the maximum nominal stress criterion [22,23], that is, interface damage begins to occur when the normal component or tangential component of the cohesive force reaches the critical value. The mathematical description of the maximum nominal stress criterion is as follows:

$$max\left\{\frac{t_n}{t_n^o}, \frac{t_s}{t_s^o}, \frac{t_t}{t_t^o}\right\} = 1, \tag{1}$$

where t_n^o, t_s^o, t_t^o correspond to the peak values of the contact stress of the interface elements in the normal direction, the first shear direction, and the second shear direction, respectively. The three parameters of t_n, t_s, t_t represent the contact stresses of the interface elements at a certain moment in the normal direction, the first shear direction, and the second shear direction, respectively. Once damage is initiated, the cohesive force gradually degrades until complete failure. The energy consumed at complete separation is the interfacial fracture energy. Referring to the fiber pull-out energy formula proposed in the literature [17, 18]:

$$\omega_{po} = \frac{\sigma_f^3 d^2 V_f}{96 l \tau_i^2}, \tag{2}$$

where l is the fiber length, σ_f is the fiber strength (taken as 2.97 GPa), τ_i is the fiber/matrix interfacial shear strength, d is the fiber diameter, and V_f is the fiber volume fraction. Substituting the relevant parameters into Equation (2), the interfacial fracture energy ω_c of the composites studied in this paper was calculated to be approximately 0.005 J/m^2.

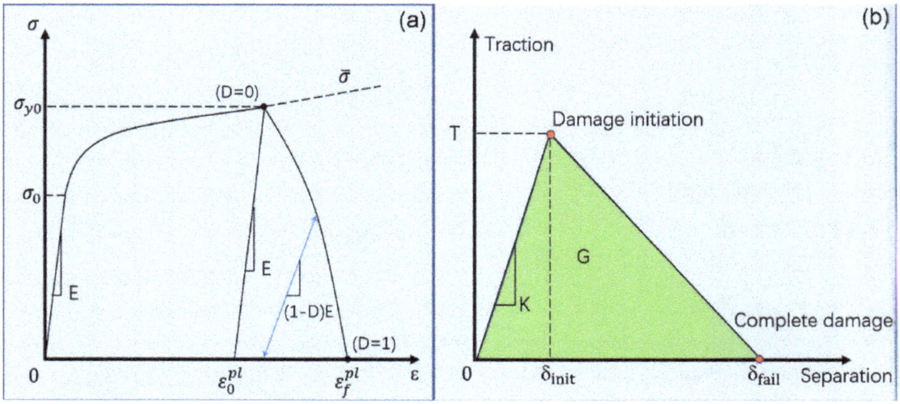

Figure 1. (**a**) Stress–strain curves with progressive damage degradation; (**b**) Bilinear traction–separation model for interface elements (The green colored part corresponds to the area under the curve).

In this work, fiber entanglement and agglomeration are not considered initially, and the fiber spacing is controlled to be one fiber diameter apart. The main reason is that when the fibers are too close to each other, the mesh generation of the model will greatly increase the computational requirements, cost, and difficulty. When the mesh density is properly designed, there is no significant impact on the results. Therefore, in this paper, six groups of predicted stress–strain curves of W_f-reinforced W-Cu composites with different mass fractions were calculated. The fiber mass fraction ranged from 1% to 6%, and the aspect ratio was 20. The composite models are shown in Figure 2a.

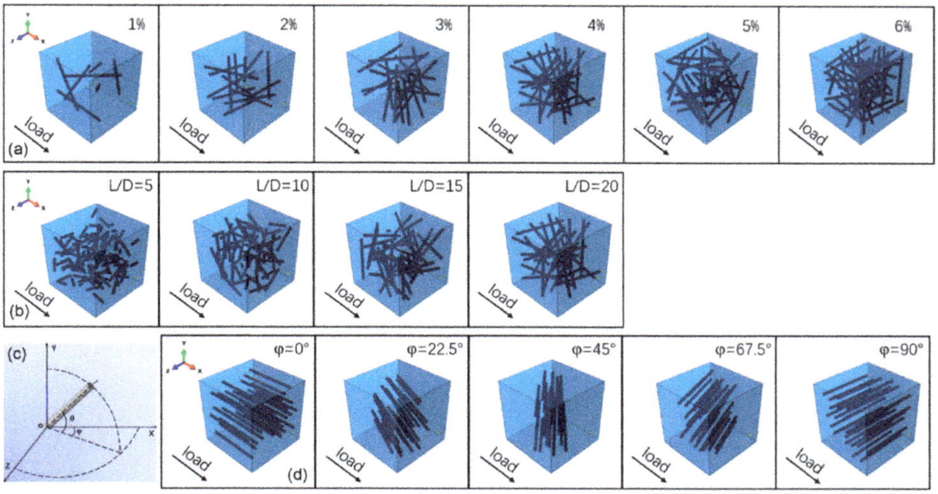

Figure 2. Microstructural mechanics models of W_f/W-Cu composites with different microstructural parameters: (**a**) fiber mass fractions; (**b**) fiber aspect ratios; (**c**) fiber orientation schematic image; (**d**) fiber orientation angles.

To investigate the influence of fiber aspect ratio on the mechanical properties of W_f/W-Cu composites, a series of microstructural mechanics models with different fiber aspect ratios were constructed in this study. The fiber diameter was fixed at 0.02 mm, and the fiber aspect ratios were set as 5, 10, 15, and 20, respectively. Figure 2b shows the representative volume elements of four randomly oriented short fiber-reinforced composites. In actual preparation processes, short fibers of different lengths can be obtained through slicing techniques. To comprehensively consider computational efficiency and result accuracy, the size of the RVE was uniformly selected as 0.5 mm × 0.5 mm × 0.5 mm. One face (X = 0) of the model was set as the fixed end of the load, and a tensile displacement load was applied to the opposite face (X = 0.5 mm). The displacement loading rate was 0.000575 mm/s. The cohesive parameters of the fiber/matrix interface were taken from the literature [24]. In this paper, t_n = 500 MPa, t_s = 200 MPa, t_t = 200 MPa, and the fracture energy was $w = 0.005\ J/m^2$.

Fiber orientation is another important factor affecting the mechanical properties of composites. To quantitatively analyze the influence of fiber orientation angle on the tensile strength of W_f/W-Cu composites, a series of microstructural mechanics models with different orientation angles (φ) were constructed based on a fixed fiber aspect ratio (L/D = 20). As shown in Figure 2c, the angle between the fiber axis and the tensile load direction is defined as φ, and the fiber axis is always parallel to the XOZ plane. φ = 0° corresponds to the fiber axis being parallel to the load direction, while φ = 90° means they are perpendicular. In this paper, five typical fiber orientation angles were considered: φ = 0°, 22.5°, 45°, 67.5°, and 90°. The corresponding microstructural models of the composites are shown in Figure 2d. The other parameters of the model were consistent with the previous text, with the fiber aspect ratio fixed at 20 and the content at 4%.

3. Results and Discussion

3.1. Investigation of the Mechanical Properties of W-Cu Composites Reinforced with Short Fibers of Different Aspect Ratios

Before analyzing the influence of fiber aspect ratio on the mechanical properties of composites, it is necessary to introduce the concept of critical fiber length. The critical length refers to the minimum fiber length required for composites to achieve fiber reinforcement. According to the shear lag theory proposed by B.W. Rosen [25], the critical fiber length *Lc* can be expressed as:

$$L_c = d_{fb}\frac{\sigma_{fu}}{2\tau_s}, \tag{3}$$

where σ_{fu} denotes ultimate strength of the fiber (MPa); d_{fb} denotes fiber diameter (mm); L_c denotes critical fiber length (mm), independent of the magnitude of the applied stress; τ_s is the shear yield stress of the matrix. When the fiber length is greater than the critical length, the fiber plays a tensile load-bearing role; otherwise, the fiber undergoes interface debonding and pullout and cannot fully exert its reinforcement effect.

For the W-Cu matrix material studied in this paper, they have a certain degree of plasticity, and the shear stress along the fiber/matrix interface is non-uniformly distributed. Figure 3 shows the distribution characteristics of fiber axial tensile stress and interfacial shear stress. It can be seen that the interfacial shear stress reaches its maximum value at the fiber ends and gradually decreases to zero along the fiber axial direction. This is mainly because of the stress concentration at the fiber ends and the additional shear stress generated by the coordinated deformation with the matrix. Taking the fiber strength as σ_{fu} = 3058 MPa and the shear yield strength of the matrix as τ_s = 200 MPa, substituting into Equation (3), it can be calculated that the critical fiber aspect ratio (L_c/d_f) is approximately 7.6. This means that when the fiber aspect ratio is greater than 7.6, the fiber can effectively play a tensile load-bearing role; otherwise, the fiber will fail due to interface debonding and pullout.

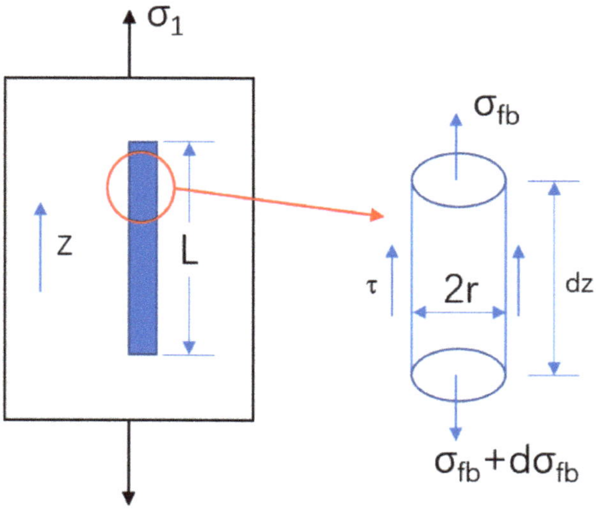

Figure 3. Stress Equilibrium of Short Fiber Microelement with Parallel External Load.

To systematically study the influence of the fiber aspect ratio on the mechanical properties of composites, a series of microstructural mechanics models of W_f/W-Cu composites with a fiber content of 4% and aspect ratios of 5, 10, 15, and 20 were constructed in this paper. Figure 4 shows the tensile stress–strain curves of composites with different aspect ratios. The results show that when the fiber aspect ratio is 5, the tensile strength of the composite is 523.9 MPa and the elongation is 5.13%, both of which are lower than other conditions. This is mainly because the fiber length is lower than the critical length at this time, and the stress concentration at the fiber ends causes plastic deformation of the matrix and induces fiber/matrix interface debonding and fiber pullout, as shown in Figure 5a. When the fiber aspect ratio increases to 10, 15, and 20, the tensile strength of the composites increases to 526.7 MPa, 533.8 MPa, and 544.2 MPa, respectively, and the elongation also correspondingly increases to 4.87%, 4.95%, and 5.05%. Figure 5b–d shows that at this time, the fiber length has exceeded the critical length, and the long fibers can effectively transfer the tensile load, and the fiber axial tensile stress gradually increases until the fiber undergoes tensile fracture. Longer fibers are conducive to stress redistribution between the fibers and the matrix, and the fibers fully exert their reinforcement effect and bear most of the external load. Comprehensive analysis shows that the fiber aspect ratio has a significant influence on the mechanical properties of composites. When the fiber length is less than the critical length (such as L/D = 5), the mechanical properties of the composite are poor; when the fiber length is greater than the critical length, as the aspect ratio increases from 10 to 20, the strength and plasticity of the composite show obvious improvements. Therefore, in the design of short fiber reinforced composites, the optimization of the fiber aspect ratio is crucial, and it should be ensured that the fiber length exceeds the critical aspect ratio to fully exert the reinforcement effect of the fibers.

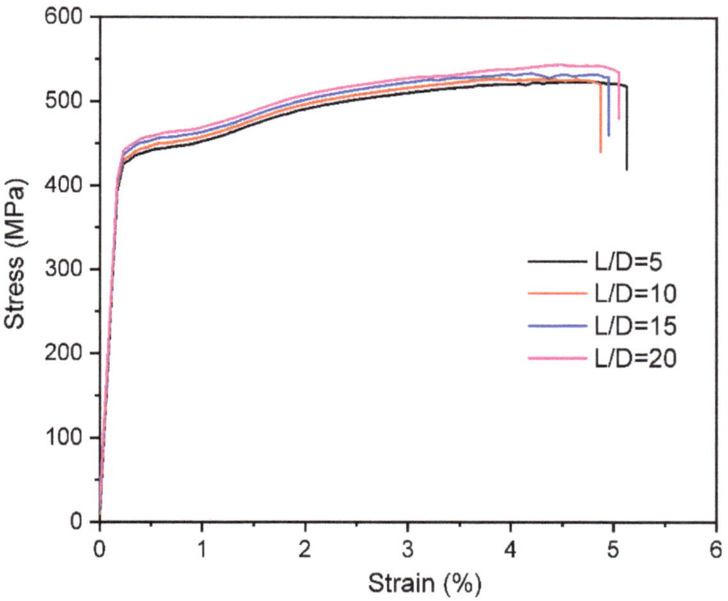

Figure 4. Tensile stress–strain curves of W_f/W-Cu composites with different fiber aspect ratios.

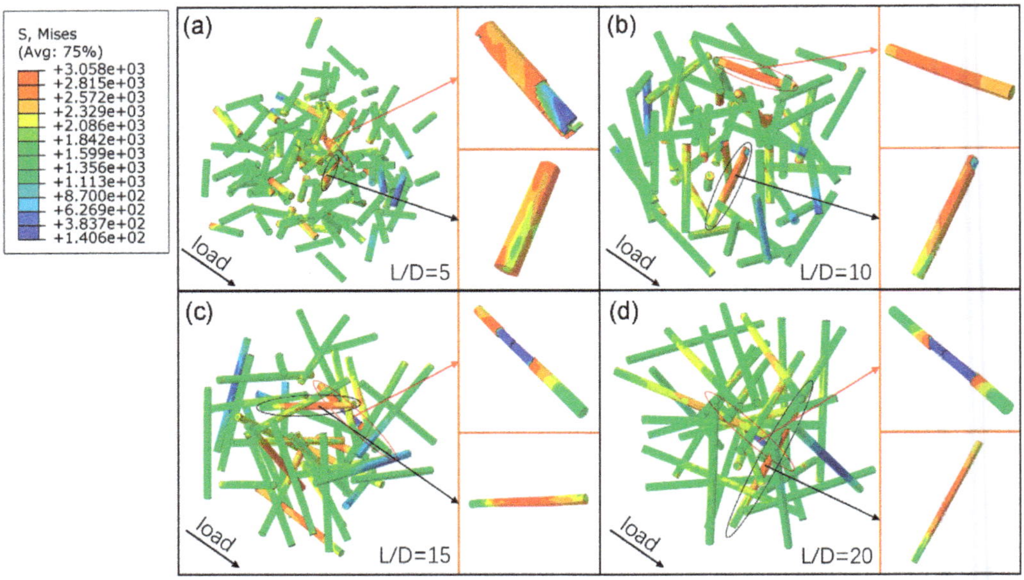

Figure 5. Contours of axial tensile stress in fibers of W_f/W-Cu composites with different fiber aspect ratios (**a**) L/D = 5, (**b**) L/D = 10, (**c**) L/D = 15, (**d**) L/D = 20.

To further reveal the influence mechanism of fiber aspect ratio on the mechanical properties of composites, Figure 6 shows the evolution of fiber Mises stress cloud diagrams of composites with different aspect ratios during the tensile deformation process. When the strain is 2.3% (Figure 6a,d,g,j), the fiber stresses in all conditions are significantly lower than their strength limit (3058 MPa), indicating that the fibers have not undergone significant damage or fracture. With further development of the deformation (strain 4.6%), the fibers begin to show local damage. For short fiber composites with an aspect ratio of 5 (Figure 6b), the fiber ends are the first to be damaged, resulting in a sudden drop in fiber stress. This is because the fiber length is short, the load transfer efficiency is low, and the fiber ends become stress concentration points. However, when the fiber length is longer (Figure 6e,h,k), the fibers can effectively transfer the load, and damage mainly occurs in the middle region of the fibers. With the fracture of local fibers, the load is distributed to adjacent intact fibers, resulting in stress adjustment. Eventually, when the strain reaches 5.75% (Figure 6c,f,i,l), most of the fibers suffer severe damage, and the composite rapidly fails. Fiber fracture induces matrix cracking, leading to macroscopic fracture of the composite.

Based on the above analysis, it can be seen that the fiber aspect ratio affects the macroscopic mechanical properties of the composite by influencing the fiber damage mode. For short fiber composites (L/D < 7.6), the fibers mainly induce matrix plastic deformation and interface debonding through end stress concentration, making it difficult to fully exert their reinforcement effect. In long fiber composites (L/D > 7.6), the fibers can effectively transfer the load, and damage mainly initiates in the middle region of the fibers. The fibers fully exert their reinforcement effect, and with the increase in aspect ratio, the strength and plasticity of the composite show significant improvements. Therefore, the rational utilization of the fiber aspect ratio effect and optimization of fiber length are crucial for improving the mechanical properties of short fiber reinforced composites.

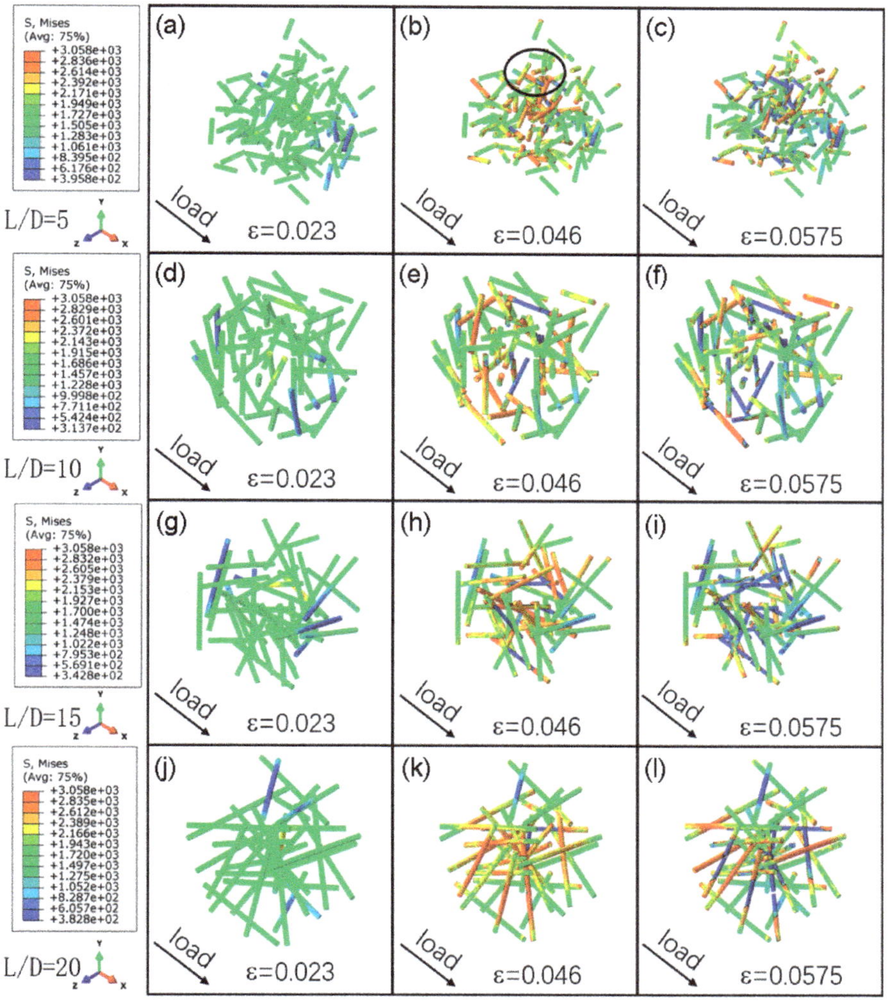

Figure 6. Evolution of Mises stress in fibers of W_f/W-Cu composites with different aspect ratios (L/D) **(a–c)** L/D = 5, **(d–f)** L/D = 10, **(g–i)** L/D = 15, **(j–l)** L/D = 20.

From the perspective of fiber reinforcement mechanisms, the strengthening effects of short fibers on metal matrix composites can be divided into the following three categories: direct load-bearing type, yield transfer type, and microcrack bridging type. For direct load-bearing type reinforcement, the fibers undergo axial tensile deformation until fiber fracture, and most of the load is borne by the fibers. In yield transfer type reinforcement, the matrix first undergoes plastic deformation and transfers the load to the fibers through shear stress until the fiber/matrix interface peels off. Microcrack bridging refers to the fibers hindering and slowing down the expansion of matrix cracks, improving the fracture toughness of the composite through the fiber bridging effect. Combining the simulation results of this paper, it can be found that fibers with different aspect ratios have different reinforcement mechanisms. Figure 7 shows the microscopic morphology of the fracture surface of W_f/W-Cu composites. When the fiber aspect ratio is less than the critical value (such as L/D = 5, Figure 7a), the fibers mainly achieve reinforcement through the pullout mechanism. Under the action of tensile load, the fiber/matrix interface first peels off, and then the fibers are pulled out of the matrix. This corresponds to the yield transfer type reinforcement mechanism, and short fibers are difficult to effectively transfer the load. As the fiber aspect ratio exceeds the critical value (L/D = 10−15, Figure 7b,c), tensile fracture of the fibers can be clearly observed, indicating that the fibers can effectively transfer the load until they fracture beyond their own strength. This corresponds to the direct load-bearing type reinforcement, which is the main strengthening mechanism of long fiber composites. When the fiber aspect ratio further increases (L/D = 20, Figure 7d), the fracture toughness of the composite is significantly improved. The bridging effect of the fibers helps to hinder crack propagation, and a large number of fibers are pulled out, corresponding to the microcrack bridging type reinforcement mechanism. In summary, the fiber aspect ratio is a key factor affecting the mechanical properties of short fiber reinforced composites. Fiber length not only affects the stress state and damage mode of the fibers but also determines the mechanism by which the fibers exert their reinforcement effect. The critical fiber length (L/D = 7.6) is an important basis for the rational design of short fiber composites. Only when the fiber length is greater than the critical length can the fibers fully exert their direct load-bearing effect. With the increase in aspect ratio (L/D = 10−20), the strength and plasticity of the composite show significant improvements. Appropriately increasing the fiber aspect ratio (L/D = 20) also helps to exert the crack bridging effect of the fibers and improve the toughness of the composite. Therefore, for practical engineering applications, a balance between strength, toughness, and processability can be achieved by optimizing the fiber aspect ratio, thereby obtaining short fiber reinforced composites with excellent mechanical properties.

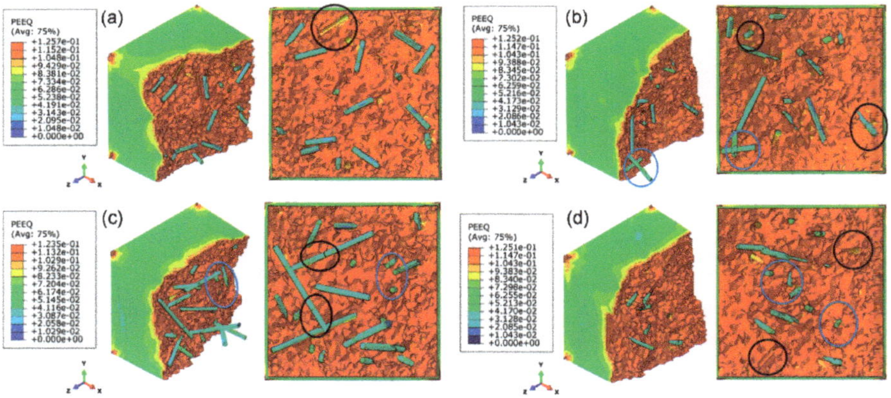

Figure 7. Fracture morphology of W_f/W-Cu composites with different fiber aspect ratios (**a**) L/D = 5, (**b**) L/D = 10, (**c**) L/D = 15, (**d**) L/D = 20.

3.2. Investigation of the Mechanical Properties of W-Cu Composites Reinforced with Short Fibers of Different Contents

In this section, the microstructural mechanics model is used to systematically study the influence of fiber mass fraction on the tensile mechanical properties of W_f/W-Cu composites. Considering the computational cost and convergence of the results, the model does not consider fiber entanglement and agglomeration phenomena, and the spacing between adjacent fibers is controlled to be above one fiber diameter. This is because excessively close fibers will severely increase the difficulty of mesh generation, leading to a sharp increase in the computational scale. Experience shows that when the mesh density is reasonable, appropriately simplifying the fiber distribution form will not have a significant impact on the computational results. Figure 8 shows the tensile stress–strain curves of composites with different fiber contents (mass fractions from 1% to 6%), where the fiber aspect ratio is fixed at 20 and randomly distributed in the matrix. All composites exhibit typical mechanical behavior of elastic deformation followed by plastic deformation until fracture. With the increase in fiber content, the tensile strength and yield strength of the composites gradually increase, while the fracture strain and elongation after fracture gradually decrease. This is mainly due to the increase in fiber content, which relatively weakens the load-bearing capacity of the matrix and decreases the plastic deformation ability of the composites.

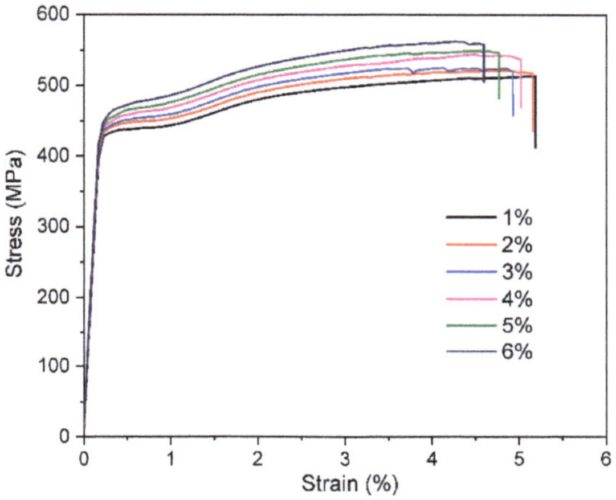

Figure 8. Tensile stress–strain curves of W_f/W-Cu composites with different mass fractions.

Quantitative analysis reveals that as the tungsten fiber content increases from 1% to 6%, the tensile strength of the composites gradually increases from 513.5 MPa to 562.5 MPa. Compared to the 1% fiber content model, the tensile strengths of the 2% to 6% models increased by 1.6%, 2.4%, 6.0%, 7.0%, and 9.6%, respectively. This result indicates that the addition of short fibers can significantly enhance the mechanical properties of the composites, and the strengthening effect increases with the increase in fiber content. At the same time, the yield strength also increases with the increase in fiber content, showing a similar trend to the tensile strength. It should be noted that the 9.55% strength improvement in our study is relatively small, mainly due to the lower reinforcement efficiency of random short fibers compared to aligned fibers. The model temporarily does not consider fiber entanglement and agglomeration, reducing fiber dispersion at high contents, and the high contribution of the plastic copper matrix to the mechanical properties. Nevertheless, our results are consistent with the expectations of classic short fiber reinforcement theory. In future research, the short fiber reinforcement effect can be maximized to obtain composites with even more excellent mechanical properties by further increasing the fiber content, optimizing the fiber dispersion and employing high-strength and high-toughness fibers and high-performance coatings. It is worth noting that the elongation after fracture of the composites generally shows a decreasing trend, but a slight rebound occurs at a fiber content of 4%. The elongation decreases from 8.26% at 2% to 7.41% at 3%, and then rebounds to 7.63% at 4%, but is still lower than the level at 2%. The reason for this phenomenon is that when the fiber content increases from 3% to 4%, the number of fibers increases significantly, and their distribution in the matrix becomes more uniform, reducing the existence of pure matrix regions. The plastic deformation of the matrix regions is the main source of elongation in the composites. A moderate increase in fiber content leads to an improvement in the uniformity of the matrix distribution, thus resulting in a slight rebound in elongation.

However, when the fiber content further increases (>4%), the constraining effect of the fibers on matrix deformation becomes dominant, and the elongation of the composites decreases again. In summary, it can be concluded that the addition of short fibers can significantly improve the mechanical properties of tungsten–copper composites, and the strengthening effect increases with the increase in fiber content. However, excessively high fiber content will lead to a decrease in the plasticity and toughness of the composites. Therefore, in practical engineering applications, a balance needs to be struck between strength, plasticity, and process cost to obtain the optimal mechanical properties.

To gain a deeper understanding of the failure mechanism of short fiber reinforced tungsten–copper composites, this paper further analyzes the damage evolution process of W_f/W-Cu composites with a fiber content of 4% under tensile loading. Figure 9 shows the equivalent plastic strain cloud diagrams and Mises stress cloud diagrams of the composites at different deformation stages. When the tensile strain reaches 4.95% (Figure 9a), some fibers have already undergone damage and fracture (red circled areas), while cracks begin to initiate in the matrix (black circled areas). High strain regions often correspond to a higher probability of matrix cracking. Figure 9d shows the corresponding Mises stress distribution cloud diagram, where the crack initiation regions also correspond to stress concentration regions, while fiber fracture leads to local unloading, and the stress level in the regions near the fibers is relatively low.

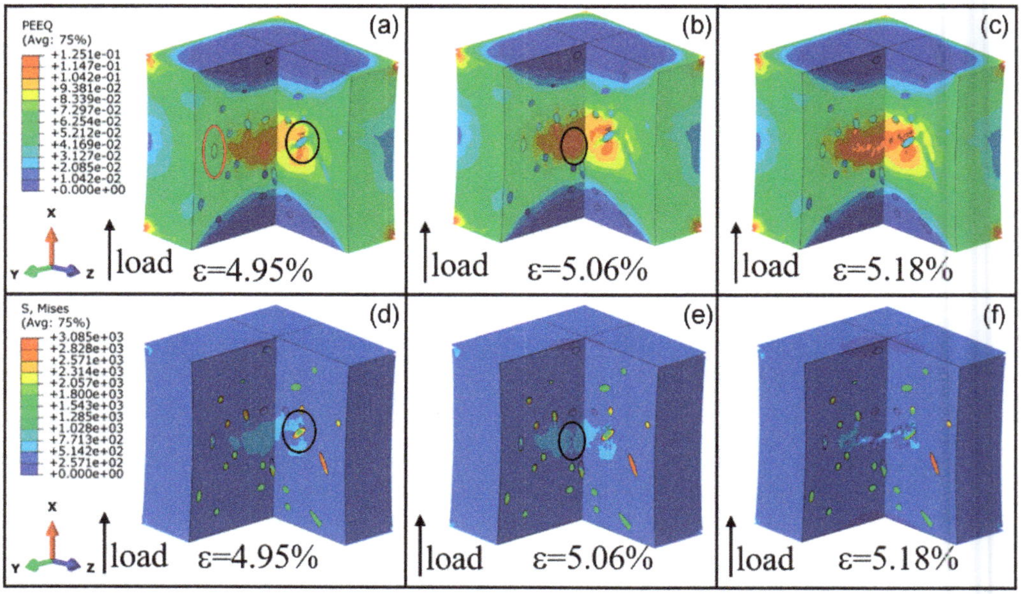

Figure 9. Equivalent plastic strain (**a–c**) and Mises stress (**d–f**) contours of Wf/W-Cu composites during tensile deformation.

With further development of the deformation (strain 5.06%, Figure 9b,e), the plastic deformation of the matrix gradually accumulates, and new crack initiation sources appear. Due to the stress redistribution caused by matrix cracking, the stress level in the regions near the crack tips is significantly reduced. At the same time, existing cracks will propagate towards the strain concentration regions, and the crack tips become the excitation points for stress concentration. Eventually, when the strain increases to 5.18% (Figure 9c,f), numerous cracks begin to intersect and coalesce, forming penetrating macroscopic fractures. The load-bearing capacity of the composites drops sharply, and the stress–strain curve decreases rapidly. The macroscopic fracture path is often formed through the interconnection of different crack tips. Figure 9c shows a typical crack propagation and coalescence path. This is mainly because the crack tip regions often correspond to high strain regions (Figure 9c) and high stress regions (Figure 9f). Propagation along the crack tip direction can effectively release the internal strain energy of the material, thus becoming the preferred direction for crack propagation.

In summary, the damage and failure process of short fiber reinforced tungsten–copper composites can be divided into three stages: (1) Fiber fracture and matrix crack initiation stage. Randomly distributed fibers first undergo damage and fracture, inducing the initiation of multiple cracks in the matrix. (2) Matrix plastic deformation and crack propagation stage. The plastic deformation of the matrix continuously accumulates, existing cracks gradually propagate, and new crack initiation sources appear. Crack tips become stress concentration points, and cracks propagate along high strain regions. (3) Macroscopic fracture formation stage. Numerous microcracks coalesce and interconnect to form penetrating macroscopic cracks, and the load-bearing capacity of the composites rapidly decreases. Cracks preferentially propagate along high strain sand high stress regions until the final failure of the material. Therefore, for short fiber reinforced composites, suppressing fiber damage and fracture and delaying matrix plastic deformation are key to improving their damage tolerance and increasing their failure toughness. Optimizing the design of the fiber/matrix interface strength and rationally controlling the fiber orientation and distribution can effectively avoid stress concentration, improve the crack propagation resistance of the composites, and thus obtain excellent mechanical properties.

3.3. Investigation of the Mechanical Properties of W-Cu Composites Reinforced with Short Fibers of Different Orientations

Figure 10 shows the tensile stress–strain curves of composites with different orientation angles. It can be seen that the fiber orientation angle has a significant influence on the mechanical behavior of the composites. When $\varphi = 0°$, i.e., the fiber axis is parallel to the load direction, the elongation after fracture of the composite reaches the maximum value of 5.2%, but the tensile strength is not the highest. This may be because parallel fibers can maximize their reinforcement effect, and the fibers undergo significant plastic deformation and interface slip, allowing the composite to exhibit better ductility.

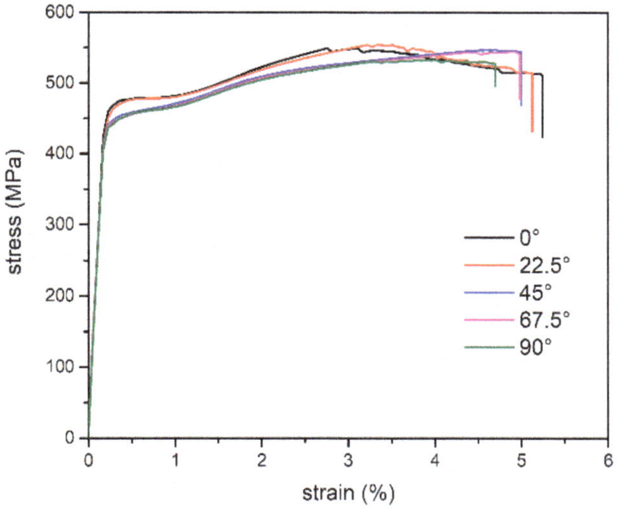

Figure 10. Tensile stress–strain curves of W_f/W-Cu composites with different fiber orientation angles.

When φ = 22.5°, the tensile strength of the composite reaches the highest value of 554.4 MPa. This indicates that moderate fiber inclination helps to exert the fiber reinforcement effect, generating additional shear stress between the fibers and the matrix, and improving the tensile resistance of the composite. As φ further increases, the tensile strength and elongation of the composite generally shows a decreasing trend. When φ = 90°, i.e., the fibers are perpendicular to the load direction, the tensile strength and elongation of the material are reduced to the lowest values of 533.2 MPa and 4.7%, respectively. This is mainly because perpendicular fibers have difficulty effectively transferring the load and are prone to peeling and brittle fracture of the fiber/matrix interface during the tensile process.

Considering both tensile strength and ductility, the tensile performance of the composite is optimal when φ = 22.5°, achieving a good balance between strength and toughness. As φ deviates from the optimal angle, the tensile performance gradually decreases, and both strength and toughness are reduced to the lowest level when φ = 90°. Therefore, in the structural design of W_f/W-Cu composites, rational control of the fiber orientation distribution is crucial. Moderate fiber inclination helps to form effective load transfer between the fibers and the matrix while leveraging the plastic deformation ability of the fibers, which is key to obtaining high-performance composites.

To further reveal the influence of fiber orientation on the failure behavior of composites, Figures 11 and 12 show the evolution processes of fiber Mises stress and equivalent plastic strain at φ = 0° and 90°, respectively. When φ = 0°, the tensile load is mainly borne by the fiber axial direction. In the early stage of deformation (ε = 2.3%, Figure 11a), the fiber stress is mainly concentrated in the middle region of the fibers, and the matrix stress level is relatively low and uniformly distributed. With further development of deformation (ε = 4.6%, Figure 11b), the middle region of the fibers begins to undergo local fracture, causing redistribution of fiber stress. Stress concentration regions begin to appear in the matrix, corresponding to high-probability regions of crack initiation. Eventually, matrix cracks connect the fiber fracture regions, forming macroscopic fractures (Figure 11c,d). Cracks preferentially propagate along the fiber alignment direction, resulting in the lowest crack propagation resistance in the matrix.

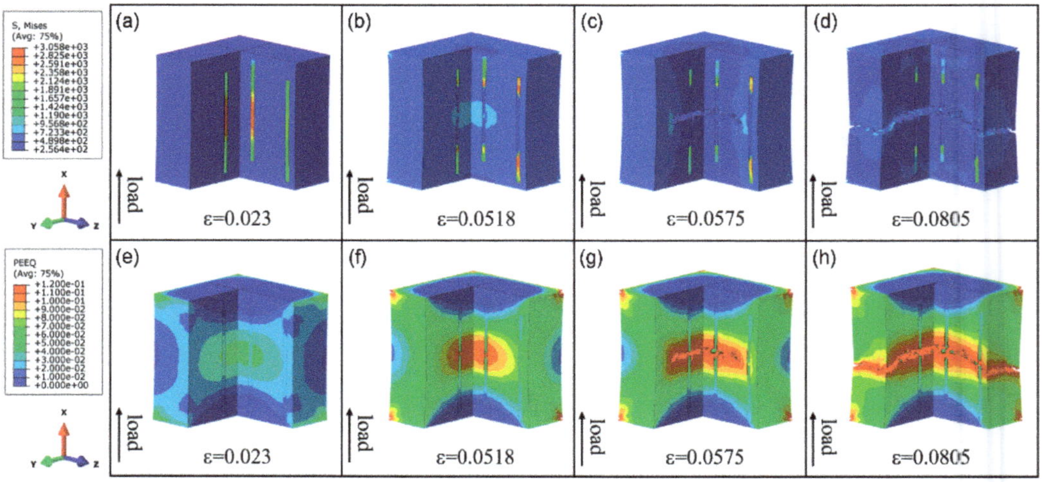

Figure 11. Evolution of Mises stress and equivalent plastic strain contours of W_f/W-Cu composites with φ = 0° (**a–d**) Mises stress; (**e–h**) Equivalent plastic strain.

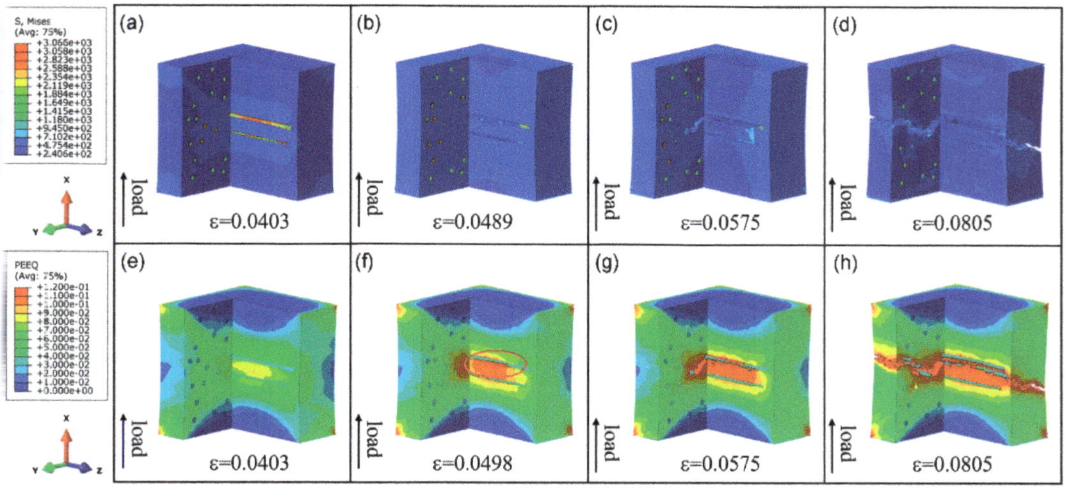

Figure 12. Evolution of Mises stress and equivalent plastic strain contours of W_f/W-Cu composites with φ = 90° (**a–d**) Mises stress; (**e–h**) Equivalent plastic strain.

When φ = 90° (Figure 12), the load is mainly transferred between the fibers and the matrix through radial tensile stress, and the load borne by the fiber axial direction is small. In the early stage of deformation (ε = 2.3%, Figure 12a), the axial distribution of fiber stress is relatively uniform. With the further development of deformation (Figure 12b–d), the fibers begin to undergo local yielding and discrete fracture. Matrix cracking is mainly controlled by the stress concentration caused by fiber radial fracture. Cracks rapidly propagate along the direction perpendicular to the fiber axis until macroscopic fracture of the material occurs. The above process fully demonstrates that when the fibers are perpendicular to the load direction, the fibers have difficulty effectively transferring the load, and the composite is more prone to brittle fracture. This is also the main reason for the significant reduction in strength and toughness of the composite when φ = 90°.

The evolution process of equivalent plastic strain in Figures 11 and 12 further reveals the regulatory effect of fiber orientation on matrix plastic deformation and crack propagation. When φ = 0°, plastic deformation is mainly limited to the fiber ends and fracture regions, and matrix deformation is relatively uniform. Cracks preferentially propagate along the fiber axial direction, corresponding to high strain regions. When φ = 90°, the fibers have difficulty constraining matrix deformation, and matrix plastic deformation becomes more significant. Cracks rapidly propagate perpendicular to the fiber direction, and the material undergoes brittle fracture along high strain bands.

In addition to affecting the macroscopic mechanical properties of composites, fiber orientation also significantly changes their microscopic fracture mechanisms. Figure 13 shows the fracture morphology of W_f/W-Cu composites with different orientation angles. When φ = 0° (Figure 13a), the fiber axis is parallel to the load direction, and the composite has relatively high tensile strength and fracture strain (Figure 10). The fracture surface is dominated by fiber tensile fracture, and the fiber fracture cross-section is flat, exhibiting cleavage fracture characteristics. This indicates that the fibers bear the main load and undergo macroscopic cleavage along the fiber fracture regions. The matrix plastic deformation induced by fiber fracture is small, and fiber pullout pits are observed only in local regions. When φ = 0°, fiber tensile fracture is the main failure mechanism of the composite.

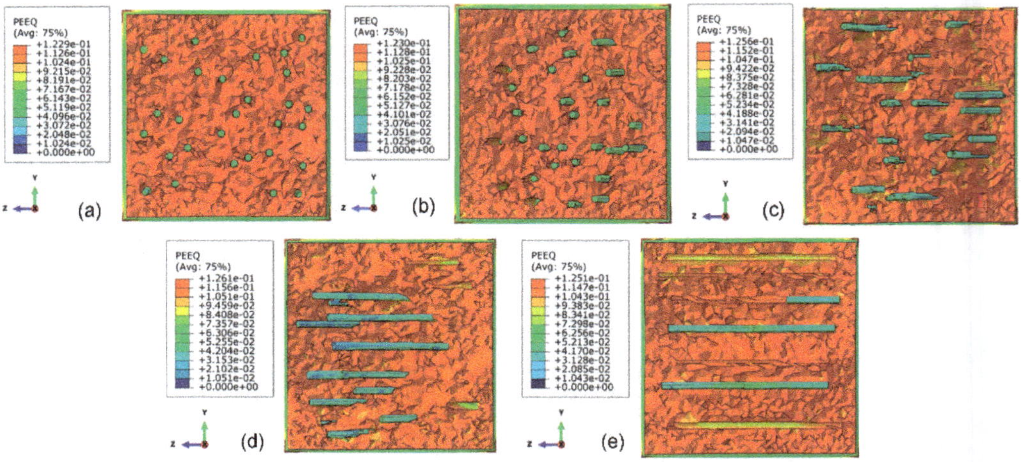

Figure 13. Fracture morphology of W_f/W-Cu composites with different fiber orientation angles φ = 0° (**a**), φ = 22.5° (**b**), φ = 45° (**c**), φ = 67.5° (**d**), φ = 90° (**e**).

When φ = 22.5° and 45° (Figure 13b,c), the composite strength reaches the highest while the plasticity slightly decreases (Figure 10). Fiber oblique fracture and local fiber pullout begin to appear on the fracture surface. As φ increases, the stress state of the fibers gradually transitions to shear loading, and fiber oblique fracture helps to improve the material toughness. At the same time, the load transfer efficiency of the fiber/matrix interface decreases, and interface debonding and fiber pullout begin to become important failure modes. Further increasing φ to 67.5° (Figure 13d), the composite strength decreases but the plasticity significantly improves. Fiber pullout becomes the main failure characteristic, with a large number of fibers pulled out from the matrix, leaving circular holes on the matrix surface. The fibers have difficulty transferring axial load, and the fiber/matrix interface first undergoes shear delamination, followed by fiber pullout, and the material fails by shear along the interface. When φ = 90° (Figure 13e), the fiber axis is perpendicular to the load direction, and both the strength and plasticity of the composite are reduced to the lowest level (Figure 10). Interfacial shear delamination dominated by fiber pullout becomes the main failure mode, and the fracture surface exhibits a honeycomb structure. Secondary shear fracture of fibers can be observed in local regions, mainly due to the compressive and shear loads perpendicular to the fiber axis during the pullout process.

Combining the mechanical property test results in Figure 10, it can be seen that fiber orientation significantly affects the microscopic fracture mechanisms of W_f/W-Cu composites. As φ increases from 0° to 90°, the tensile strength of the composite first increases and then decreases, the fracture toughness first decreases and then increases, and the fracture mechanisms successively undergo three stages of fiber tensile fracture, fiber shear yielding, and fiber pullout: (1) When φ = 0°, fiber tensile fracture dominates, and both the strength and toughness of the material are relatively high. (2) When φ = 22.5°, fiber tensile fracture and shear yielding coexist, the strength reaches the highest while the toughness slightly decreases. (3) When φ = 45° to 67.5°, fiber oblique fracture and local pullout coexist, the strength decreases but the toughness significantly improves. (4) When φ = 90°, fiber pullout dominates, and both strength and toughness are reduced to the lowest level. This indicates that fiber orientation significantly changes the microscopic fracture mechanisms of composites by influencing fiber/matrix load transfer, fiber deformation mode, and interface debonding. In the design of short fiber composites, the fiber orientation distribution can be optimized to balance various failure modes such as fiber tension, shear, and pullout, achieving the best match between strength and toughness, and obtaining composites with excellent comprehensive mechanical properties. Based on the systematic RVE simulations, this study has made a preliminary attempt at a numerical analysis of macroscopic mechanics, but the current macroscopic model is still relatively simplified and needs further improvement and modification. On the one hand, it is necessary to develop a more refined constitutive model to better reflect the anisotropy and nonlinear characteristics of the composite material. On the other hand, it is necessary to correct the key parameters in the model through more comprehensive experimental characterization and verification (such as fracture analysis, fatigue experiments, etc.) to improve the accuracy and applicability of the predictions. This will be the focus of our next step of research.

To summarize, based on physical mechanisms, a multi-scale mechanical constitutive model was constructed to quantitatively analyze the influence of fiber parameters such as content, aspect ratio, and orientation on the mechanical properties of composites, revealing the synergistic reinforcement mechanism of short fibers. However, the current research mainly focuses on theoretical simulations and lacks direct comparison and validation with experimental results. In subsequent research, we will focus on strengthening work in both simulation and experimentation. On the one hand, we will further improve the microstructural mechanics model, considering the influence of practical factors such as fiber entanglement and agglomeration, to improve the accuracy and applicability of the model. On the other hand, we will obtain performance data of composites with different fiber parameters through mechanical property testing experiments such as tensile, compression, and fracture toughness tests, and systematically compare them with simulation results. The close integration of simulation and experimentation can more comprehensively and deeply reveal the intrinsic correlation between fiber parameters and material mechanical behavior, providing theoretical and experimental basis for composition optimization and structural design of W_f/W-Cu composites.

4. Conclusions

In this paper, a three-dimensional microstructural mechanic model containing the fiber/matrix interface was constructed to systematically investigate the influence of aspect ratio (5–20), fiber content (1–6%), orientation (0–90°), and other factors on the mechanical properties and failure mechanisms of the material. The main conclusions are as follows:

(1) There exists a critical value (7.6) for the fiber aspect ratio. When below this value, fiber pullout dominates, and when above this value, fiber tensile fracture is the main mechanism, corresponding to yield transfer type interface strengthening and load transfer type direct reinforcement, respectively. Rational matching of short/long fibers can significantly improve the comprehensive mechanical properties of composites.

(2) As the fiber content increases from 1% to 6%, the tensile strength of the composite increases by 9.6%, the yield strength increases by 10.2%, while the elongation after fracture decreases by 18.6%, indicating that the short fiber content significantly affects the strength and plasticity of the composite. Excessively high fiber content leads to increased fiber yielding and fracture, weakening the deformation ability of the matrix.

(3) As the fiber orientation angle increases from 0° to 90°, the tensile strength of the composite first increases and then decreases, and the fracture toughness first decreases and then increases. The strength is highest when $\varphi = 22.5°$, and the toughness is best when $\varphi = 0°$. Fiber orientation regulates load transfer, fiber deformation mode, and interface debonding, significantly changing the microscopic fracture mechanisms of the material.

(4) The multi-scale damage evolution mechanism of short fiber composites is revealed: fiber fracture induces matrix crack initiation, followed by matrix plastic deformation,

crack propagation, and interconnection, ultimately leading to macroscopic fracture. The higher the fiber content, the more severe the fiber yielding and fracture; the closer the fiber orientation is to perpendicular, the more significant the interfacial shear damage.

(5) This research innovatively constructs a three-dimensional microstructural mechanic model, achieves high-fidelity simulations of the entire deformation and damage process of short fiber composites, and quantitatively reveals the multi-scale synergistic mechanism of fiber parameters regulating the mechanical behavior of the composite. The research results can provide an important theoretical basis for the structural design and performance optimization of short fiber reinforced composites. The constitutive model and multi-scale analysis method established in this study can also serve as a reference for studying the mechanical properties of other short fiber composite systems. Future research can investigate the influence of fiber spatial distribution, interfacial chemical bonding, and other factors on the material behavior, further improve the multi-scale simulation method, and provide theoretical guidance for the engineering application of short fiber composites in extreme environments.

Author Contributions: L.Z. conceived the idea, led the methodology, investigation, organization, and supervised the work, also responsible for writing the original draft, reviewing and editing the manuscript, and acquiring funding. Y.X. contributed significantly to the investigation, methodology development, and participated in writing the original draft. B.L. and N.L. assisted with the experimental work, data collection, and participated in the manuscript's review and editing process. H.X. was involved in data analysis, visualization, and contributed to writing, reviewing, and editing the manuscript. B.C. and H.W. provided valuable insights and suggestions throughout the study and contributed to reviewing and editing the final manuscript. All authors have read and agreed to the published version of the manuscript.

Funding: This research was funded by National Natural Science Foundation of China grant number [51604223] And The APC was funded by National Natural Science Foundation of China grant number [51604223].

Data Availability Statement: The original contributions presented in this study are included in the article. Further inquiries can be directed to the corresponding author(s).

Acknowledgments: This work was financially supported by the National Natural Science Foundation of China (51604223).

Conflicts of Interest: The authors declare no conflict of interest.

References

1. Sharma, S.K.; Saxena, K.K.; Salem, K.H.; Mohammed, K.A.; Singh, R.; Prakash, C. Effects of various fabrication techniques on the mechanical characteristics of metal matrix composites: A review. *Adv. Mater. Process. Technol.* **2024**, *10*, 277–294. [CrossRef]
2. Singh, B.; Kumar, I.; Saxena, K.K.; Mohammed, K.A.; Ijaz Khan, M.; Ben Moussa, S.; Arif, M.; Pruncu, C.I.; Misra, R.D.K. A future prospects and current scenario of aluminium metal matrix composites characteristics. *Alex. Eng. J.* **2023**, *76*, 1–17. [CrossRef]
3. Samal, P.; Vundavilli, P.R.; Meher, A.; Mahapatra, M.M. Recent progress in aluminum metal matrix composites: A review on processing, mechanical and wear properties. *J. Manuf. Process.* **2020**, *59*, 131–152. [CrossRef]
4. Zhuo, L.; Liu, C.; Yin, E.; Zhao, Z.; Pang, S. Low-cost and low-temperature 3D printing for refractory composite inspired by fused deposition modeling and binder jetting. *Compos. Part A Appl. Sci. Manuf.* **2022**, *162*, 107147. [CrossRef]
5. Zhuo, L.; Zhao, Z.; Qin, Z.; Chen, Q.; Liang, S.; Yang, X.; Chai, Z.; Pan, W.; Zhang, G. Enhanced mechanical and arc erosion resistant properties by homogenously precipitated nanocrystalline fcc-Nb in the hierarchical W-Nb-Cu composite. *Compos. Part B Eng.* **2019**, *161*, 336–343. [CrossRef]
6. Wang, Y.L.; Zhuo, L.C.; Yin, E.H. Progress, challenges and potentials/trends of tungsten-copper (W-Cu) composites/pseudo-alloys: Fabrication, regulation and application. *Inter. J. Refract. Met. Hard Mater.* **2021**, *100*, 105648. [CrossRef]
7. Lorenzin, G.; Jeurgens, L.P.H.; Cancellieri, C. Stress tuning in sputter-grown Cu and W films for Cu/W nanomultilayer design. *J. Appl. Phys.* **2022**, *131*, 225304. [CrossRef]

8. Kelly, A.; Tyson, W.R. Tensile properties of fibre-reinforced metals: Copper/tungsten and copper/molybdenum. *J. Mech. Phys. Solids* **1965**, *13*, 329–350. [CrossRef]
9. Lee, W.; Sue, W.; Lin, C. The effects of temperature and strain rate on the properties of carbon-fiber-reinforced 7075 aluminum alloy metal-matrix composite. *Compos. Sci. Technol.* **2000**, *60*, 1975–1983. [CrossRef]
10. Halpin, J.C.; Kardos, J.L. The Halpin-Tsai equations: A review. *Polym. Eng. Sci.* **1976**, *16*, 344–352.
11. Zeng, T.; Zhang, R.; Schumacher, G. Effect of fiber orientation on the tensile properties of carbon fibers reinforced magnesium matrix composites. *Mater. Charact.* **2021**, *172*, 110871.
12. Shirvanimoghaddam, K.; Hamim, S.U.; Karbalaei Akbari, M.; Fakhrhoseini, S.M.; Khayyam, H.; Pakseresht, A.H.; Ghasali, E.; Zabet, M.; Munir, K.S.; Jia, S.; et al. Carbon fiber reinforced metal matrix composites: Fabrication processes and properties. *Compos. Part A Appl. Sci. Manuf.* **2017**, *92*, 70–96. [CrossRef]
13. Chao, X.; Qi, L.; Ma, W.; Ge, J.; Tian, W. Monte carlo based strategy to simulate the microstructure evolution of the short-fiber reinforced metal matrix composites. *Mater. Today Commun.* **2022**, *33*, 104275. [CrossRef]
14. Thomason, J.L.; Vlug, M.A. Influence of fibre length and concentration on the properties of glass fibre-reinforced polypropylene: 1. Tensile and flexural modulus. *Compos. Part A Appl. Sci. Manuf.* **1996**, *27*, 477–484. [CrossRef]
15. Bahl, S. Fiber reinforced metal matrix composites—A review. *Mater. Today Proc.* **2021**, *39*, 317–323. [CrossRef]
16. Kelly, A. *Strong Solids*; Clarendon Press: Oxford, UK, 1973.
17. Zhuo, L.; Luo, B.; Zhao, Z.; Zhang, Y.; Zhang, J.; Liang, S.; Cao, L.; Chai, Z.; Pan, W.; Zhang, G. Infiltrated tungsten-copper composite reinforced with short tungsten fibers. *Vacuum* **2020**, *173*, 109123. [CrossRef]
18. Mileiko, S.T.; Serebryakov, A.V.; Kolchin, A.A. Mechanics of composites with short fibers. *Mech. Compos Mater.* **2002**, *38*, 255–264.
19. Hull, D.; Clyne, T.W. *An Introduction to Composite Materials*; Cambridge University Press: Cambridge, UK, 1996.
20. Sim, J.; Yeo, D.; Yoon, H.; Yu, S.; Lee, D.; Bae, J. Experimental and Digimat-FE Based Representative Volume Element Analysis of Dye-Mixed Colored Resin and Carbon Fiber. *Polymers* **2022**, *14*, 1028. [CrossRef]
21. Balasubramani, N.K.; Zhang, B.; Chowdhury, N.T.; Mukkavilli, A.; Suter, M.; Pearce, G.M. Micro-mechanical analysis on random RVE size and shape in multiscale finite element modelling of unidirectional FRP composites. *Compos. Struct.* **2022**, *282*, 115081. [CrossRef]
22. Kim, J.; Mai, Y. High strength, high fracture toughness fibre composites with interface control—A review. *Compos. Sci. Technol.* **1991**, *41*, 333–378. [CrossRef]
23. Yue, C.Y.; Looi, H.C.; Quek, M.Y. Assessment of fibre-matrix adhesion and interfacial properties using the pull-out test. *Int. J. Adhes. Adhes.* **1995**, *15*, 73–80. [CrossRef]
24. Tian, W.; Qi, L.; Su, C.; Zhou, J.; Jing, Z. Numerical simulation on elastic properties of short-fiber-reinforced metal matrix composites: Effect of fiber orientation. *Compos. Struct.* **2016**, *152*, 408–417. [CrossRef]
25. Rosen, B.W. Tensile failure of fibrous composites. *AIAA J.* **1964**, *2*, 1985–1991. [CrossRef]

crystals MDPI

Review

Progress in the Copper-Based Diamond Composites for Thermal Conductivity Applications

Kang Chen, Xuesong Leng *, Rui Zhao, Yiyao Kang and Hongsheng Chen *

Institute of Special Environments Physical Sciences, Harbin Institute of Technology (Shenzhen), Shenzhen 518055, China
* Correspondence: lengxuesong@hit.edu.cn (X.L.); chenhongsheng@hit.edu.cn (H.C.)

Abstract: Copper-based diamond composites have been the focus of many investigations for higher thermal conductivity applications. However, the natural non-wetting behavior between diamond particles and copper matrix makes it difficult to fabricate copper-based diamond composites with high thermal conductivity. Thus, to promote wettability between copper and diamond particles, the copper/diamond interface must be modified by coating alloying elements on the diamond surface or by adding active alloying elements with carbon in the copper matrix. In this paper, we review the research progress on copper-based diamond composites, including theoretical models for calculating the thermal conductivity and the effect of process parameters on the thermal conductivity of copper-based diamond composites. The factors that affect interfacial thermal conductivity are emphatically analyzed in this review. Finally, the current problems of copper-based diamond composites and future research trends are recommended.

Keywords: copper-based diamond composite; thermal conductivity; interface; alloying elements; surface treatment

1. Introduction

Electronic devices generate heat which is increasing due to the continuous development of miniaturization and integration of microelectronics [1]. Excessive heat deteriorates the long-term reliability and performance of electronic devices, and conventional thermal management materials are no longer adequate for dissipating heat with the rapid development of electronic products. Thus, there is a significant need to improve heat dissipation in electronic devices, and therefore, there is an urgent need for materials with high thermal conductivity to meet contemporary thermal requirements. The thermal conductivity of traditional electronic packaging substrates (crystalline silicon) is 120–150 W/(m·K) [2]. Diamond has extremely high thermal conductivity of 1200~2500 W/(m·K), and therefore, it has attracted attention for its excellent thermal properties [3–9]. However, the fabrication process of diamond on electronic devices is very difficult and expensive, which makes it difficult for diamond to be used directly in the electronics industry. There is another solution that involves combining diamond particles and copper matrix (copper has a thermal conductivity of 400 W/(m·K). The surface tension of copper at its melting point is between 1000 and 1500 mN/m, and the melting point of copper is 1083.4 °C with diamond particles, which not only reduces production costs, but also results in a composite with high thermal conductivity. However, due to the high surface tension of copper, the wetting angle between copper and diamond is large. Moreover, diamond and copper are difficult to chemically react with each other, ultimately leading to poor wetting between copper and diamond. The poor wettability between diamond and copper can promote the formation of defects at the interface and weaken the interfacial bonding between diamond and copper, finally, leading to lower thermal conductivity of copper-based diamond composites [10]. At high temperatures, diamond transforms into a more stable graphite, and this phenomenon causes a decrease in thermal conductivity. Therefore, many researchers have given massive attention to methods of mitigating the problems mentioned above.

Until recently, two main methods had been reported in the literature for improving the thermal conductivity of copper-based diamond composites. One method is diamond surface metallization and the other method is matrix alloying. Diamond surface metallization involves coating a strong carbide-forming element on the diamond surface by using electroplating, magnetron sputtering, salt bath coating, or vacuum micro evaporation plating. Additionally, matrix alloying involves adding strong carbide-forming elements into copper matrix, which includes gas atomization and alloy melting. The purpose of both methods is to form several layers of carbides on the diamond surface during the fabricating processes. The formed carbides can improve the wettability between the diamond particles and the copper matrix, thus, increasing the thermal conductivity of copper-based diamond composites. Various methods have been used to prepare copper-based diamond composites, such as high-temperature and high-pressure sintering (HTHP) [11], vacuum hot press sintering (VHPS) [12], spark plasma sintering (SPS) [13], and melt infiltration [14]. All these methods aim at increasing the density and wettability of copper-based diamond composites by applying external conditions.

Theoretical models have been developed and have emerged as powerful tools for calculating the thermal conductivity of copper-based diamond composites. Although several studies have shown that the thermal conductivity calculated by using theoretical models differs considerably from the measured experimental data, theoretical models can reasonably reflect the influence of several parameters, such as diamond volume fraction, thermal conductivity of diamond and copper, diamond particle size, diamond morphology, and interfacial thermal resistance, on the thermal conductivity of copper-based diamond composites. In the following sections, we discuss the effects of these parameters in theoretical models on the thermal conductivity of copper-based diamond composites.

2. Theoretical Models

Many researchers have proposed theoretical models for calculating the thermal conductivity of copper-based diamond composites. Theoretical models can promote a better understanding of the relationships between the thermophysical properties of copper-based diamond composites and their process parameters, and can provide relatively accurate predictions of the thermal conductivity of copper-based diamond composites. Several models are widely used, including the Maxwell–Eucken model, the Hasselman–Johnson model, and the differential effective medium model. In this section, we provide brief descriptions of these theoretical models.

2.1. Maxwell–Eucken Model

The Maxwell–Eucken model [15] was the first theoretical model applied to predict the thermal conductivity of copper-based diamond composites. It can be described by the following equation [15]:

$$\lambda_c = \lambda_m \frac{2(1 - V_d)\lambda_m + (1 + 2V_d)\lambda_d}{(2 + V_d)\lambda_m + (1 - V_d)\lambda_d} \quad (1)$$

where λ_c is the thermal conductivity of copper-based diamond composites, λ_m is the thermal conductivity of copper matrix, λ_d is the thermal conductivity of the diamond reinforcing particles, and V_d is the volume fraction of diamond spherical particles in the copper-based diamond composites.

The Maxwell–Eucken model assumes that the diamond reinforcing particles are spherical and uniformly distributed in the continuous matrix phase and that there is no relative interaction among the diamond particles in a homogeneous copper matrix. However, this model is limited in its ability to accurately predict the thermal conductivity of composites with a low volume fraction of diamond reinforcing particles. In addition, when the diamond volume fraction exceeds 20%, the theoretical values differ significantly from experimental results, possibly due to the interaction between particles and the defects presented in the composites under normal conditions. One way to improve the agreement between theoretical and experimental values is to improve the wettability between the diamond particles and copper matrix.

2.2. Hasselman–Johnson Model

The thermal conductivity of particle-reinforced composites can be measured more accurately by using the Hasselman–Johnson model, which was developed on the basis of the Maxwell theoretical model [16]. The Hasselman–Johnson model is widely used to predict and analyze the thermal conductivity of particle-reinforced composites. The equation for the Hasselman–Johnson model is given as [16]:

$$\lambda_c = \lambda_m \frac{[2(\frac{\lambda_d}{\lambda_m} - \frac{\lambda_d}{ah_c} - 1)V_d + \frac{\lambda_d}{\lambda_m} + 2\frac{\lambda_d}{ah_c} + 2]}{[(1 - \frac{\lambda_d}{\lambda_m} + \frac{\lambda_d}{ah_c})V_d + \frac{\lambda_d}{\lambda_m} + 2\frac{\lambda_d}{ah_c} + 2]} \quad (2)$$

where λ_c represents the thermal conductivity of the copper-based diamond composite; λ_m and λ_d are the thermal conductivity of the copper matrix and the diamond reinforcing particles, respectively; V_d is the diamond volume fraction in the copper-based diamond composite; a is the radius of the diamond reinforcing particles; h_c is the interfacial thermal conductivity that is determined by the matrix and the reinforcing phase.

As the diamond volume fraction increases, the assumptions made in the Hasselman–Johnson model, such as the assumption of spherical particles and the neglect of particle clustering or agglomeration, become less accurate. These limitations lead to an overestimation of the thermal conductivity at higher diamond volume fractions, which is not consistent with experimental data. Therefore, the Hasselman–Johnson model is more reliable and applicable at lower diamond volume fractions in copper-based diamond composites.

2.3. Differential Effective Medium Model

Bruggeman developed a differential effective medium model. He assumed that particle content grows slowly throughout the embedding process. The equation for thermal conductivity was derived from the Bruggeman integral embedding principle [17], which takes into account the forces present in the particles when the volume fraction of particles is too large, and can be described as [17]:

$$(1 - V_d) = \frac{(\lambda_c - \lambda_d^{eff})}{(\lambda_m - \lambda_d^{eff})} \times \left(\frac{\lambda_m}{\lambda_c}\right)^{\frac{1}{3}} \quad (3)$$

where λ_m is the matrix thermal conductivity and λ_d^{eff} is the effective thermal conductivity of the diamond particles, which can be calculated by using the following equation:

$$\lambda_d^{eff} = \frac{\lambda_d}{1 + \frac{\lambda_d}{ah_c}} \quad (4)$$

where a is the radius of the diamond particles. The trend of the interfacial thermal conductivity is consistent with the trend of the thermal conductivity of the composite. Since V_d, λ_d, λ_m, and a are the same for all samples, the interfacial thermal conductivity (h_c) dominates the composite thermal conductivity (λ_c), to a large extent, and λ_c varies with the penetration time.

To better study the gap between theoretical models and actual data, we conducted the following work. We set the thermal conductivity of diamond to 2000 W/(m·K) and the thermal conductivity of copper to 400 W/(m·K). Similarly, we assumed an interfacial thermal resistance of 2.9×10^7 W/(m^2·K) (experimentally determined by Y. Zhang [18]). By varying the diamond volume fraction, we obtained the thermal conductivity of different theoretical models, which are plotted in Figure 1a. Figure 1a also compares the theoretical models with the experimental data of Q.L. Che [19]. In the theoretical models, the calculated thermal conductivities increase as the diamond volume fraction increases. In reality, the thermal conductivity first increases, and then decreases with an increase in the diamond volume fraction. The reasons for this phenomenon are discussed in detail in the following section.

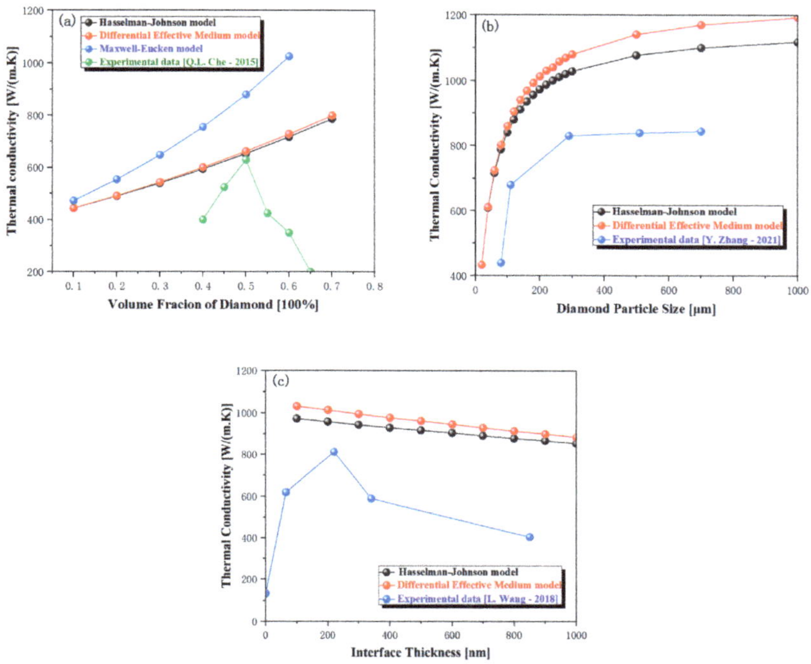

Figure 1. Comparison of thermal conductivity calculated by different models with experimental data: (**a**) Volume fraction of diamond; (**b**) diamond particle size; (**c**) interface thickness. The experimental data in Fig. 1a are from Ref. [19]; the experimental data in Fig. 1b are from Ref. [20]; the experimental data in Fig. 1c are from Ref. [21].

We assumed that the volume fraction of diamond is 68.7%. Different theoretical models were employed to calculate the thermal conductivity of different sized diamond particles. Comparing these thermal conductivity values with Y. Zhang's data [20], we obtained Figure 1b. It can be seen that the thermal conductivity of copper-based diamond composites gradually increases as the diamond particle size increases. However, the effect of diamond particle size on thermal conductivity is gradually weakened as the diamond particle size.becomes larger We can also see that the trend of the experimental data is consistent with the trend of the theoretical models. The reasons for this phenomenon are discussed in detail in the following section.

To analyze the relationship between interfacial thermal conductivity and thermal conductivity of copper-based diamond composites in theoretical models, it is necessary to understand the factors that affect interfacial thermal resistance of diamond particles. The total interfacial thermal resistance can be obtained by calculating the interfacial thermal resistance of each layer, as shown in Figure 2 [22]. The equation for total interfacial thermal resistance is described as [22]:

$$R_{\text{total}} = R_{\text{Cu/carbide}} + l_{\text{carbide}}/k_{\text{carbide}} + R_{\text{diamond/Carbide}} \tag{5}$$

where R_{total} is the total interfacial thermal resistance between the copper matrix and the diamond particles; $R_{Cu/carbide}$ is the interfacial thermal resistance between the copper matrix and the carbide layer; $l_{carbide}$ is the interface thickness and $k_{carbide}$ is the thermal conductivity of the interface; $R_{diamond/carbide}$ is the interfacial thermal resistance between the diamond particles and the carbide layer. It should be noted that thermal resistance is equal to the reciprocal of thermal conductivity.

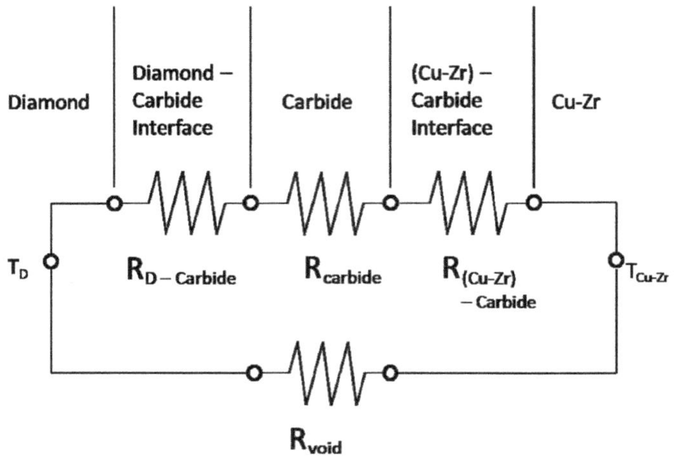

Figure 2. Thermal circuit of the carbide interface in the copper-based diamond composites. Copyright 2015, with permission from Elsevier [22].

To establish the interface thermal conductivity of the interface without considering its actual microstructure, an acoustic mismatch model can be used [23,24]:

$$h = 1/4\rho_{in}C_{in}v_{in}\eta \quad (6)$$

where ρ is the density of the material; C is the specific heat; v is the phonon velocity determined from the longitudinal and transverse velocities, the subscript indicates the incident side of the phonon; η is the average probability of phonon transport across the interface. The experimental data and the required parameters can be found in Y. Zhang's data [18]. Subsequently, we can calculate $h_{Cu/carbide}$ and $h_{diamond/carbide}$. Therefore, the total effect of interfacial thermal conductivity on the thermal conductivity of copper-based diamond composites can be considered to be the effect of the carbide interfacial thickness on the thermal conductivity of copper-based diamond composites. The relationship between carbide interface thickness and the thermal conductivity of copper-based diamond composites can be derived, as shown in Figure 1c. At the same time, we compared the thermal conductivity obtained from the theoretical models with the experimental data from L. Wang [21]. In the theoretical models, the thermal conductivity decreases as the interface thickness increases. In reality, the thermal conductivity first increases, and then decreases as the interface thickness increases. The reasons for this phenomenon are discussed in detail in the following section.

Observing Figure 1, we can easily observe that the thermal conductivity calculated by the theoretical models is higher than the actual experimental data. The existing theoretical models do not consider enough factors that affect the thermal conductivity, such as interface defects, compactness, and other problems. Therefore, there is a significant difference between theoretical values and experimental values.

The above three models have a wide range of applications. Additionally, more theoretical models have been proposed. For example, the Hamilton–Grosser model takes the shape of the diamond reinforcing particles into account. The Agari model considers the heat transfer chain formed by the interactions among particles as well as the shapes and relative content of the particles, etc. In addition to theoretical models, researchers have recently used molecular dynamics simulation to predict the thermal conductivity of composite material interfaces. Y. Zhang [25] used molecular dynamics to simulate the effect of interface structure on the thermal conductivity of composite materials. The results indicated that the G value of the W_2C/diamond interface was significantly higher than that of the Cu/diamond and W/diamond interfaces. Additionally, W_2C was capable of reducing the vibrational misfit between copper and diamond. The purpose of these models is to obtain theoretical values that are as close as possible to experimental data. However, due to the complexity of the factors that affect thermal conductivity, many parameters need to be considered, such as the thermal conductivity of particles and matrix, particle size and shape, particle content, defects, particle interactions, etc. However, none of the existing models can fully match the heat transfer mechanism for the thermal conductivity of composite materials. Thus, further research needs to be conducted to modify these theoretical models.

3. The Effect of Diamond Particle Size

Diamond particle size is an essential factor that affects the thermal conductivity of copper-based diamond composites, as highlighted in both the Hasselman–Johnson model and the differential effective medium model. An increase in diamond particle size leads to an increase in the thermal conductivity of copper-based diamond composites, as observed in Figure 1.

Consequently, the use of nanoscale diamond as raw material could result in low thermal conductivity of copper-based diamond composites. For instance, in a study conducted by S.V. Kidalov [26], nanodiamonds were compared to natural micro diamonds (5~7 μm and 10~14 μm) in the production of sintered copper-based diamond composites, using a pressure of approximately 6.0~6.5 GPa and sintering temperatures of 1000~2000 °C, for 6~20 s. The experimental results showed that the thermal conductivity of the nanodiamond composite was less than 10 W/(m·K), which was significantly lower than that of the natural micro diamond composite (500 W/(m·K)). Figure 3 depicts the effect of nanodiamond content on the thermal conductivity of copper-based diamond composites [26]. Several researchers have investigated the cause of this phenomenon. The reduction in thermal conductivity of nanodiamond composites might be attributed to the graphitization of diamond. In a study conducted by S.V. Kidalov [26], it was noted that the nanodiamonds changed from their original white color to a light grey color as the temperature increased to 1100 °C, indicating that diamond, an isomer of carbon, was stable at room temperature and atmospheric pressure but underwent transformation into graphite when heated above 800 °C.

It is true that increasing diamond particle size can increase the thermal conductivity of copper-based diamond composites [27,28]. K. Hanada [27] investigated the effect of varying diamond grain size (0.005~7.7 μm) on the thermal conductivity of copper-based diamond composites by using a powder metallurgical process. They found that when the diamond volume fraction was controlled at 1%, the thermal conductivity of copper-based diamond composites increased as the diamond particle size increased, and the diamond particles were uniformly distributed in the copper matrix. Similarly, Y. Wu [28] investigated the effect of diamond particle size (10~400 μm) on the microstructure, interface, and thermal conductivity of copper-based diamond composites. The obtained results are shown in Figure 4, where it can be seen from this figure that the thermal conductivity of copper-based diamond composites increases with an increase in diamond particle size [28].

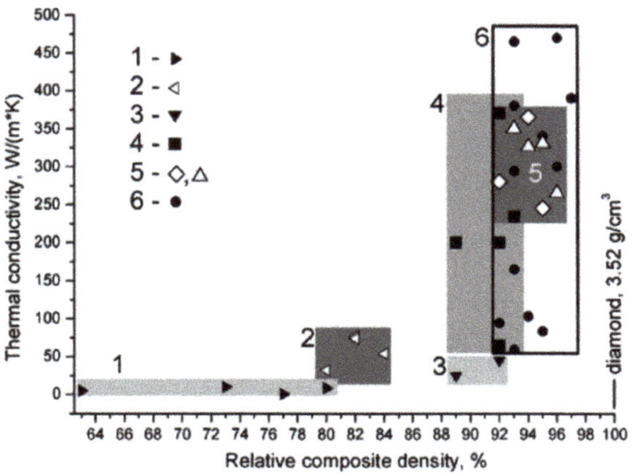

Figure 3. Thermal conductivity of copper-based diamond composites with different diamond size combinations. The composite densities are normalized to that of single crystal diamond. 1—100% nanodiamonds; 2—50% 10~14 μm diamonds and 50% nanodiamonds; 3—85% 10~14 μm diamonds and 15% nanodiamonds; 4—100% 5~7 μm diamonds; 5—99.5% 10~14 μm diamonds and 0.5% fullerenes combined with data on 99% 10~14 μm diamonds and 1% nanodiamonds; 6—100% 10~14 μm diamonds. Copyright 2007, with permission from Elsevier [26].

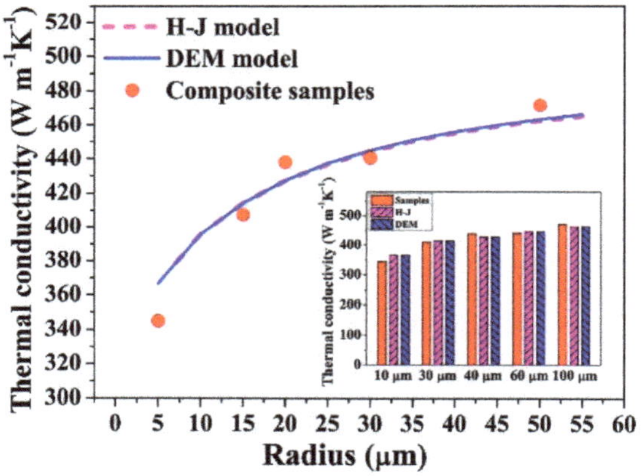

Figure 4. Thermal conductivities of copper-based diamond composites with different sizes of diamond particles. Copyright 2019, with permission from Elsevier [28].

Y. Zhang [20] discovered that with smaller sized diamond particles, many sharp polygonal carbides were formed on the (100) surface of diamond particles, as shown in Figure 5b. However, when the diamond particle size was large enough, the diamond surface was uniformly covered with carbides, as depicted in Figure 5c. The (100) surface of diamond particles was formed by the two-fold bonding of carbon atoms and the surface energy of the (100) plane was 9.207 J/m^2, while the (111) surface was created by three-fold bonding, with a lower surface energy of 3.387 J/m^2. Carbides preferentially formed on the (100) surfaces with higher surface energy when the amount of alloying elements in the copper matrix was insufficient [20]. As diamond particle size increased, the area of the interface decreased and carbides were more likely to spread evenly over the diamond surface. Consequently, copper-based diamond composites with larger sized diamond particles have higher densities and higher thermal conductivity.

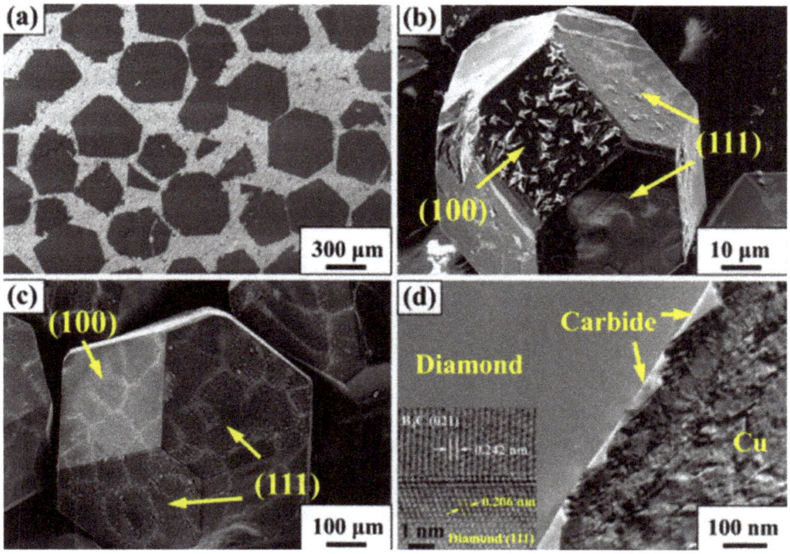

Figure 5. Microstructures of copper-based diamond composites by alloying B element in copper matrix: (**a**) The backscattered electron image; the extracted diamond particles with particle sizes of (**b**) 66 μm and (**c**) 701 μm; (**d**) the copper/diamond interface image. Copyright 2021, with permission from Elsevier [20].

An increase in diamond particle size results an increase in thermal conductivity of copper-based diamond composites, as inferred from the above content. Figure 6 displays the thermal conductivity of copper-based diamond composite materials that have been researched in recent years, and includes the highest thermal conductivity values obtained by various methods. This figure includes results from studies, in the literature, that have been conducted in the past five years and investigated thermal conductivity exceeding 600 W/(m·K). Diamond/Cu-0.5 wt% Zr has the highest thermal conductivity among composites containing Zr element. Diamond/Cu-0.3 wt% B has the highest thermal conductivity among composites containing B element. The diamond(Cr)/Cu-0.5 wt% Cr composite material has the highest thermal conductivity among those containing chromium element. It can be observed that the result is consistent with the above conclusion. However, in recent years, the size of diamond particles has basically not exceeded 300 μm. One reason for this is that larger diamond particle sizes are more expensive. Another possible reason is that the research on copper-based diamond composite materials has become more comprehensive, which has led to better bonding between diamond particles and copper at the interface. Although increasing the size of diamond particles can still enhance thermal conductivity, the effect is not as significant as previous studies suggested. Therefore, the diamond particle size employed by recent studies in the literature has mostly been between 100 μm and 300 μm.

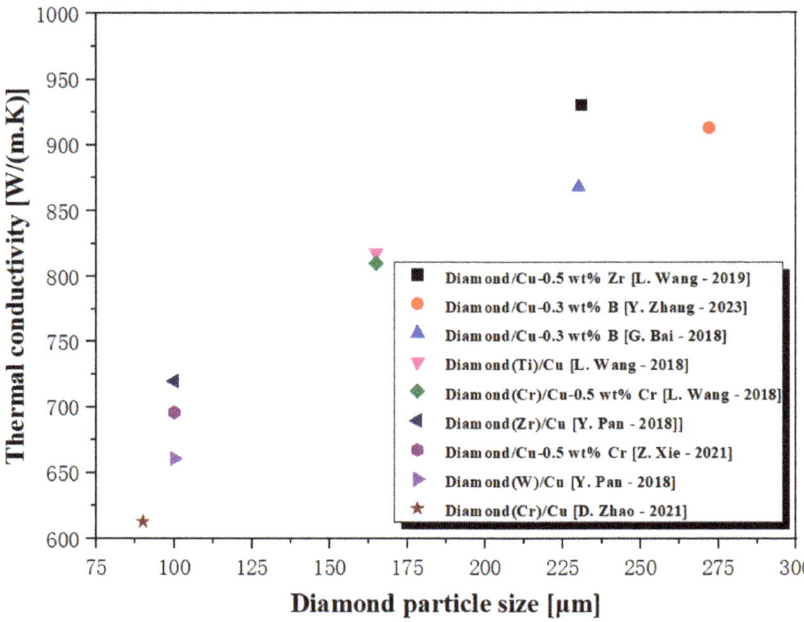

Figure 6. Comparison of thermal conductivity values from recent studies in the literatures [21,29–36].

4. Effect of Diamond Volume Fraction

The volume fraction of diamond is a vital parameter in all the theoretical models. Therefore, it is necessary to obtain a general range through extensive experiments even if the optimum diamond volume fraction is not entirely consistent for different process parameters.

According to the relationship between diamond volume fraction and thermal conductivity of copper-based diamond composites in the theoretical models, it can be concluded that with an increase in diamond volume fraction, the thermal conductivity of copper-based diamond composites increases as well. However, some studies have suggested a different conclusion [9,37]. Y. J. Zhang [9] found that by simulating the thermal conductivity of copper-based diamond composites there were minor distinctions in thermal conductivity numerical simulation results when the diamond volume fraction was less than 40%. However, when the diamond volume fraction was more than 40%, the thermal conductivity of copper-based diamond composites decreased, which was contrary to the numerical simulation results. J. Tao [37] used spark plasma sintering (SPS) to prepare copper-based diamond composites with different diamond volume fractions (50%, 60%, and 70%). As shown in Figure 7, with an increase in the diamond volume fraction, the relative density and the homogeneity of copper-based diamond composites decreased [37]. These studies showed that a larger diamond volume fraction may cause side effects. The reason for this phenomenon is that when the volume fraction of diamond is large, the copper can hardly fill the gaps among the diamond particles. As a result, the density of copper-based diamond composites decreases. The thermal conductivity also deteriorates with the diamond volume fraction. The relationship between diamond volume fraction and thermal conductivity can be summarized as followings: With increasing diamond volume fractions, the thermal conductivity of copper-based diamond composites initially increases, and then decreases.

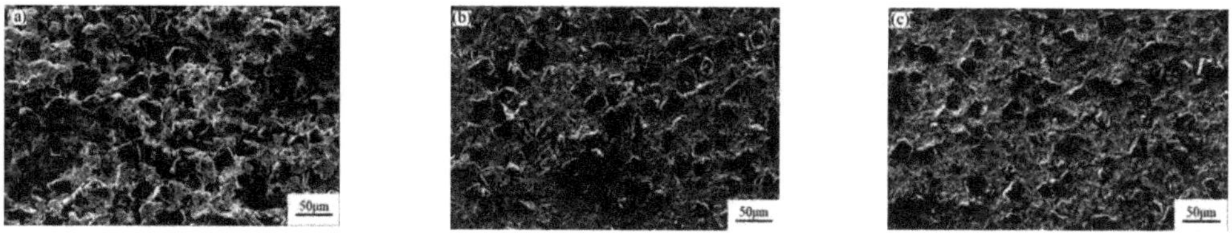

Figure 7. Microstructures of copper-based diamond composites with different diamond volume fractions sintered at 750 °C: (**a**) 70%; (**b**) 60%; (**c**) 50%. Copyright 2014, with permission from Elsevier [37].

Typically, to obtain copper-based diamond composites with high thermal conductivity, the diamond volume fraction should be less than 70%. However, high thermal conductivity can also be obtained at ultra-high pressures, even if the diamond volume fraction exceeds 70%. J. He [38] prepared copper matrix with zirconium-added composites reinforced with approximately 90 vol% of diamond particles at high temperature and pressure. According to these results, the matrix successfully filled the gaps among diamond particles. However, an improvement in thermal conductivity was not obvious and the cost was too high. For the preparation of copper-based diamond composites with effective cost and higher thermal conductivity, a diamond volume fraction between 50 and ~60% is recommended.

In addition to thermal conductivity, the coefficient of thermal expansion is also a key indicator. The volume fraction of the reinforcement is a key factor affecting the coefficient of thermal expansion of composite materials. Since the intrinsic coefficient of thermal expansion of diamond is low, increasing the volume fraction of diamond can reduce the coefficient of thermal expansion. Z. Xie [39] used a vacuum pressure infiltration method to prepare coper-based diamond composites. By measuring the CTE of composites with different diamond volume fractions, it was shown that CTE decreased with an increase in the diamond volume fraction.

Research on diamond volume fraction is quite mature. Due to variations in process parameters, there is not a unique optimal value for diamond volume fraction. In recent years, the range of diamond volume fractions used at high temperature and high pressure tends to fall between 50 and 70%. Under ultra-high pressure, the diamond volume fraction can exceed 70% or even reach 90%. Therefore, without new shaping methods, it will be difficult to achieve breakthroughs in the study of diamond volume fraction in copper-based diamond composite materials.

5. Factors Affecting Interfacial Thermal Conductivity

Improving interfacial thermal conductivity is a critical issue in copper-based diamond composites. Poor wettability between copper and diamond particles results in the low thermal conductivity of copper-based diamond composites. To enhance the interfacial thermal conductivity of copper-based diamond composites, alloying elements can be added to ameliorate the wettability between copper and diamond particles [40–43]. The type and content of alloying elements can affect the interfacial thermal conductivity, and different process parameters can also affect the thermal conductivity of copper-based diamond composites. In the following sections, we discuss methods to improve the interface thermal conductivity of copper-based diamond composites and we analyze the effect of parameters on the interface thermal conductivity.

5.1. Methods for Improving Wettability

The wettability between diamond particles and copper is very poor. Thus, by simply mixing diamond particles and copper, a composite with relatively low density and thermal conductivity is usually obtained. There are two methods to improve the wettability between diamond particles and copper. One method is pre-metalizing the diamond surface, which is coating strong carbide-forming elements on the diamond surface. The other option is pre-alloying the copper matrix, which is adding carbide-forming elements into the copper matrix prior to preparing the copper-based diamond composites.

5.1.1. Diamond Surface Treatment

To improve interfacial thermal conductivity, a diamond surface pretreatment is required. A diamond surface pretreatment refers to the removal of impurities left on the diamond surface or the application of a special treatment before the coating diamond surface, because impurities may cause the coating to fail. Cleaning, catalysis, etching, sensitizing, and activating diamond surfaces are the most common treatment methods. J. Sang [44] proposed a new pretreatment process for the preparation of copper-based diamond composites, which involved bombarding diamond particles with an argon ion beam before depositing a tungsten coating on the diamond particles. The prepared copper-based diamond composite had good interfacial bonding and the composite interface was free of micro seams and cracks. In 2019, X. Wu [45] first etched diamond particles with molten salt to obtain diamond particles with different surface roughness. Then, tungsten was coated on the sample followed by copper. Compared to the unetched copper-based diamond composite, the etched copper-based diamond composite had a 12% increase in thermal conductivity. In 2021, X. Wu [46] prepared composites with roughened surfaces using vacuum hot-press sintering (the diamond was previously etched with molten salt and the content was 38 vol.%). After molten salt etching, the thermal conductivity of the diamond/Cu composite increased significantly from 347 W/(m·K) to 433 W/(m·K) (an increase of approximately 25%). In addition, this value was further increased to 512 W/(m·K) by the addition of titanium (approximately 27% increase).

Layers of reactive elements coated on diamond surface should be able to form stable chemical bonds with the carbon elements on the diamond surface. These reactive elements contain Ti, Cr, W, Mo, B, etc. Researchers have experimentally verified that surface treatment of diamond can indeed improve the thermal conductivity of copper-based diamond composites. The methods of diamond surface metallization generally include electroplating, magnetron sputtering coating, salt bath coating, vacuum micro-evaporation coating, etc. In the following sections, we discuss the above methods in more detail.

Electroplating (EP) is the process of depositing metal on a surface, which is plated by a reduction reaction in the absence of an external power source and the presence of a strong reducing agent. Through electroplating, metal can be deposited on diamonds in a physical encapsulation process that creates no metallurgical bonds. Therefore, to improve the density of copper-based diamond composites, the diamond surface is usually coated with the active elements by other methods, and then coated with copper by electroplating. C. Zhang [47] prepared copper-based diamond composites with high densities and good thermal properties by designing a bilayer structure on diamond particles. The diamond surface was first coated with tungsten, and then coated with copper. H. Li [48] used a molten salt containing NaCl, KCl, and NaF to form a thin TiC coating on diamond, followed by chemical copper plating. The double coating (TiC-Cu) improved the thermal performance of diamond/Cu composite materials. The TC of diamond (60 vol%)/Cu composite material reached 495.5 W/(m·K), which was much higher than Cu, and slightly lower than the values predicted by the H–J model and DEM. Y. Pan [33] prepared copper-based diamond composites with good thermal conductivity using another method. They coated tungsten as the inner layer using salt bath plating, and then coated copper as the outer layer using chemical plating. As can be seen from Figure 8, the copper layer obtained by simply plating copper on the diamond surface tended to peel off [33]. In the case of first plating tungsten, the copper adhered more easily to the diamond surfaces. However, the method has several disadvantages, for example, this method requires a high external surface for the diamond particles. Thus, the diamond surface of specimens must be pretreated as mentioned above to ensure a better bond between the diamond and the plating layer. Another problem is that the use of strong reducing agents promotes graphitization of diamond.

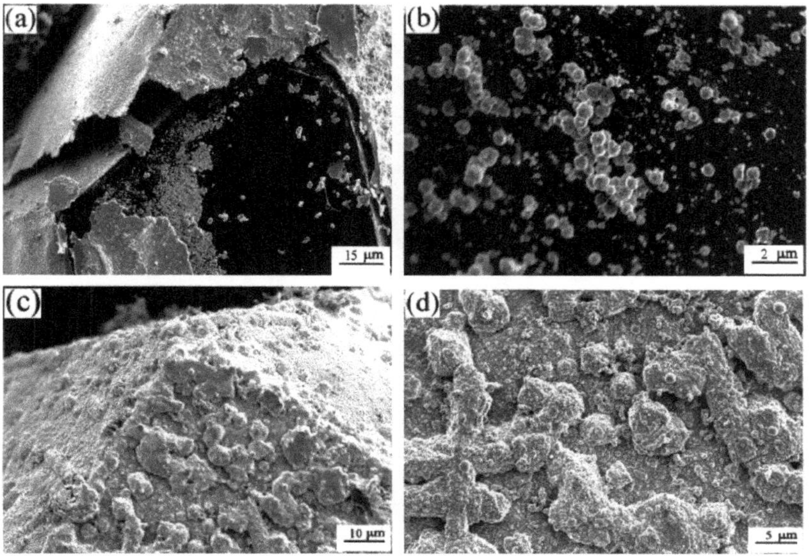

Figure 8. Microstructures of the plated copper on diamond surfaces: (**a**,**b**) Copper on raw diamond surface; (**c**,**d**) copper on tungsten-coated diamond surface. Copyright 2018, with permission from Elsevier [33].

Magnetron sputtering (MS) is a process that involves adding a small amount of inert gas, such as argon, in a vacuum environment. In this environment, the inert gas collides with electrons and produces ionization under an electric field. Then, the resulting ionophores are accelerated by using an electric field and directed toward a metal target, which sputters out target atoms and deposits them onto the diamond surface. MS can obtain a uniform and controllable coating thickness that can effectively control the thermal resistance of the interface. J. Sang [49] prepared WC layers with a thickness of 70~400 nm on diamond substrates, and obtained a maximum thermal conductivity of 943 W/(m·K) for the copper-based diamond composite. As shown in Figure 9, the plating obtained at different sputtering times was continuous and homogeneous, which was beneficial for improving the thermal conductivity of the copper-based diamond composite [49]. J. Jia [50] deposited tungsten coatings with a thickness of 45~300 nm on the surface of diamond particles by using the MS method. They found that tungsten coatings with a thickness of above 93 nm were the most effective in improving the thermal conductivity of a copper-based diamond composite and reducing the thermal resistance at the interface between diamond and copper. However, the MS method has some limitations, for example, the equipment is expensive and the single plating volume is low, resulting in high manufacturing costs, which make this method unfavorable for industrial production.

Salt bath coating (SBC) is a process in which metal powder and diamond are heated in molten salt. Diamond reacts with the molten metal at high temperatures under vacuum or protective atmosphere conditions, resulting in the formation of a coating. This method can prepare copper-based diamond composites with good interfacial bonding and a significant increase in thermal conductivity. SBC is a highly efficient, cost-effective, and simple method, which has a wide range of applications. H. Li [48] used molten salt to form a thin coating of TiC on diamond particles, followed by chemical plating of copper. As the temperature increased, the thickness of the formed TiC coating increased, resulting in cracking and flaking at the interface. The copper coating effectively covered the TiC/diamond by using chemical plating, resulting in a final thermal conductivity of copper-based diamond composite of 495.5 W/(m·K). Q. Kang [51] used a reaction medium of WO_3 in a mixed molten NaCl-KCl salt to form a WC coating on diamond surface. Copper-based diamond composites were prepared by using pressure infiltration with a vacuum, resulting in a higher thermal conductivity of 658 W/(m·K). Q. Kang [52] obtained diamond with a chromium coating from the reaction of chromium mixed in molten salt, and prepared a copper-based diamond composite with thermal conductivity of 562 W/(m·K). Although a Cr_7C_3 coating was applied to the diamond particles, the diamond was easily graphitized under this method and the separating process of diamond particles was complicated.

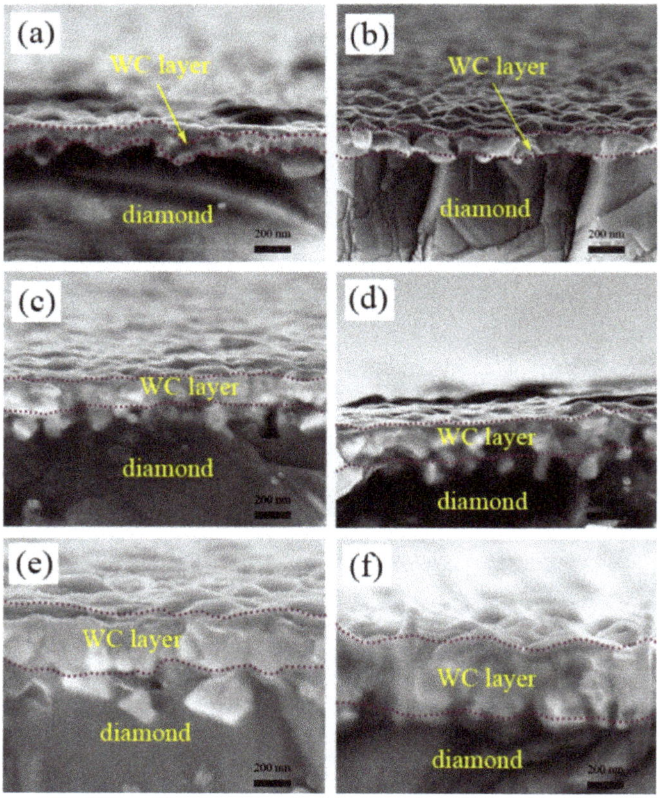

Figure 9. Cross-section microstructures of the WC layers with different thicknesses on diamond plates: (**a**) 70 nm; (**b**) 117 nm; (**c**) 179 nm; (**d**) 231 nm; (**e**) 320 nm; (**f**) 400 nm. Copyright 2022, with permission from Elsevier [49].

Vacuum micro-evaporation plating (VMEP) is a process that activates the atoms on the surface of the metal under vacuum conditions. The activated atoms diffuse to the diamond surface and react with the diamond partiles to form a film layer. The advantages of VMEP include simple processing conditions, high-quality film formation, and minimal damage to the diamond particles. L. Wang [32] prepared Cr-coated diamond particles by vacuum evaporation deposition in a thermal evaporation system. Copper-based diamond composites with a thermal conductivity of 810 W/(m·K) were prepared by using the pressure infiltration method. The disadvantages of VMEP are the complexity of the equipment and the high level of resources required to maintain the equipment each year, resulting in higher manufacturing costs.

5.1.2. Matrix Alloying

Doping copper matrix with alloying elements such as Ti, B, Cr, and Zr through pre-alloyed treatments can enhance the thermal conductivity of copper-based diamond composites. This treatment forms a carbide layer on the diamond surface during the manufacturing processes and improves the thermal conductivity of copper-based diamond composites. The copper matrix has a smaller wetting angle with carbides than diamond, resulting in higher density and better thermal conductivity of copper-based diamond composites. Matrix alloying methods usually include gas atomization and alloy melting.

Gas atomization is a widely used method to prepare copper alloy powders. It involves crushing a liquid metal or alloy into small droplets by using high-pressure air, which then rapidly condenses into a powder. Although gas atomization offers mature processing, low costs, and a high powder formation rate, the complex structure of nozzles demands higher process requirements.

Alloy smelting involves adding metals and alloying elements to a heating furnace, melting them, and then cooling to form an alloy matrix. This method is usually employed to prepare bulk copper alloys and is commonly used with liquid phase infiltration. However, to obtain copper-based diamond composites with high thermal conductivity, pneumatic infiltration is the only viable method. Y. Zhang [30] developed a Cu-B/diamond composite material by using a homemade pneumatic infiltration device. The diamond particles were densely packed in graphite molds, and a Cu-0.3 wt% B alloy ingot was placed on top. The ingots were heated to 1423 K and maintained for 30 min before applying a gas pressure of 1.5 MPa for penetration. As a result, the copper-based diamond composite exhibited a thermal conductivity of 913 W/(m·K). Figure 10 indicates that B element is mainly concentrated at the interface in the copper-based diamond composite obtained by this method, with only a very small amount of B element left in the copper matrix [30]. The enrichment of B element at the interface reduces the negative effect of alloying elements on the thermal conductivity of the copper matrix.

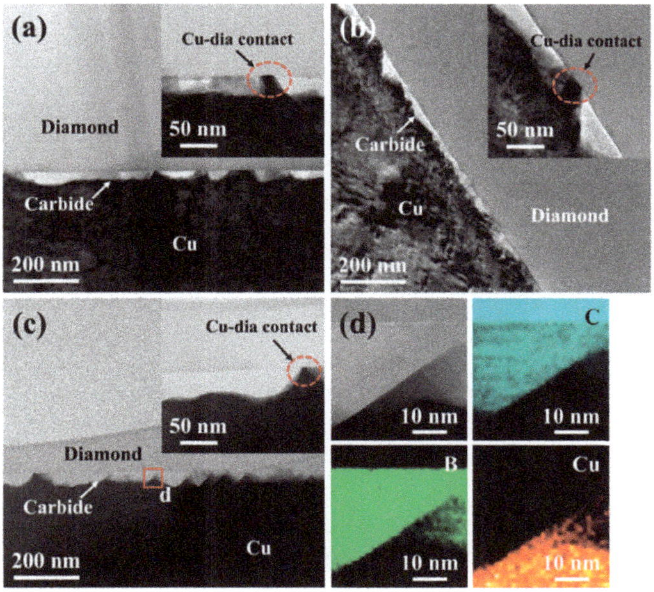

Figure 10. Interfacial microstructures of Cu-B/diamond composite with different infiltration contact times of: (**a**) 5 min; (**b**) 30 min; (**c**) 90 min. (**d**) Enlarged microstructure and mapping images in the square in (c). Copyright 2023, with permission from Elsevier [30].

In addition to adding the elements mentioned above, we can also improve thermal conductivity by adding rare earth elements. X. Zhang [13] mixed diamond with copper powder, chromium powder, and tantalum oxide powder. Using the spark plasma sintering (SPS) technique, they obtained a diamond (60 vol%)/copper (40 vol%) composite material with a thermal conductivity of 872 W/(m·K). The results showed that a well-arranged and evenly distributed interface chromium carbide layer (thickness ranging from 5 to 20 nm) was formed in the interfacial region. This phenomenon played a critical role in increasing the active interfacial binding amount and reducing defects. The addition of Sc_2O_3 increased the number of heat transfer carriers at the interface, making the interface energy system more stable. The obtained diamond/copper composite showed an effectively improved wettability between diamond particles and copper. Rare earth elements are a good choice as reinforcing materials in composite materials for the following reasons: Adding an appropriate amount of rare earth elements can refine the grain size and reduce the wetting angle between reactants, and it can also change the shape and distribution of impurities. At room temperature, rare earth elemens are almost insoluble in copper and have little effect on the thermal conductivity of the copper matrix. Rare earth elements are surface-active elements that can improve the fluidity of the metal matrix by reducing the surface tension of the metal matrix. The solubility of rare earth metals in the metal matrix is relatively low, making it easy to adsorb at the phase boundary. This fills the interface defects and also enhances the interface energy between the reinforcing particles and the metal matrix [53].

5.2. Effect of Interface

The interface between diamond and copper is a crucial aspect of copper-based diamond composites. However, the natural non-wetting behavior between diamond and copper makes it challenging to achieve copper-based diamond composites with high thermal conductivity. To improve the thermal conductivity of copper-based diamond composites, reactive elements are added to the copper matrix to form carbides, which minimize the resistance of phonons passing through the carbide layer. The thermal resistance of interface is given by Equation (3). The theoretical model suggests that as the interface thickness increases, the thermal conductivity of copper-based diamond composites should decrease. However, experimental evidence has shown that this is not always the case. Z. Xie [35] synthesized copper-based diamond composites by using Cu-Cr alloys containing 0.3–1.0 wt% Cr mixed with diamond. Copper-based diamond composites with interfacial thicknesses ranging from 80 nm to 1100 nm between copper and diamond were obtained. Figure 11 shows that the interface between copper and diamond is thin and not very dense when the Cr content is 0.3 wt%, resulting in direct contact between diamond and copper, and creating a discontinuous interface. When the Cr content is 0.5 wt%, the interface thickness is 440 nm, and the interface is thin and homogeneous [35]. However, when the Cr content is too high, significant cracks appear at the interface. The reason for the cracks is that when the interface thickness is too large, cracks appear between the diamond and carbide during high-temperature cooling due to the different expansion coefficients of diamond and carbide. Therefore, to obtain copper-based diamond composites with high thermal conductivity, the interface thickness must be well controlled.

The interface between copper and diamond is also a key factor affecting the coefficient of thermal expansion. The goal of reducing the interface coefficient of thermal expansion can be achieved by enhancing the strength of the copper and diamond interface. Q. Kang [54] obtained copper-based diamond composites using pure copper-coated diamond particles (with a WC coating on the surface) through vacuum pressure infiltration. The thermal conductivity and thermal expansion behavior of the obtained copper-based diamond composites were studied. The intermediate layer played a positive role in limiting copper's thermal expansion due to the strong bonding between the copper matrix and diamond. Generally speaking, modifying the copper/diamond interface to promote interface bonding is beneficial for reducing the coefficient of thermal expansion.

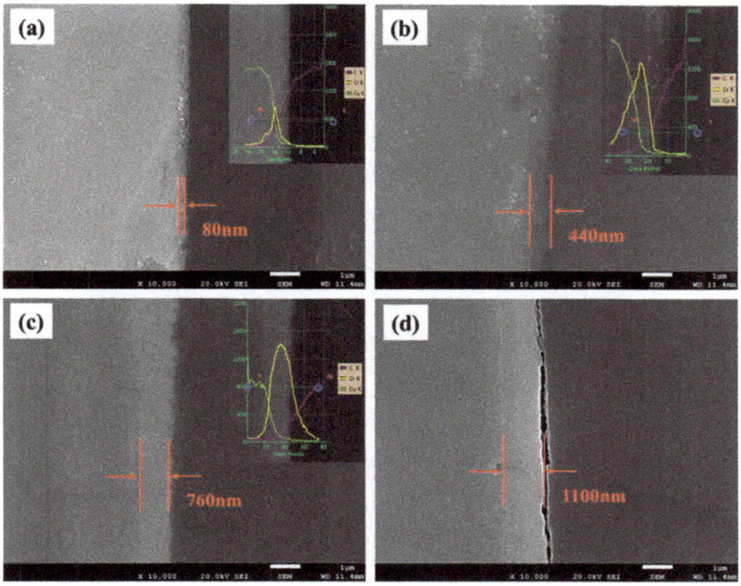

Figure 11. The interface microstructures and corresponding line scanning of copper-based diamond composites with different Cr contents: (**a**) 0.3 wt% Cr; (**b**) 0.5 wt% Cr; (**c**) 0.7 wt% Cr; (**d**) 1.0 wt% Cr. Copyright 2021, with permission from Elsevier [35].

In early studies of the copper/diamond composite material interface, the research on monolayer films is quite mature. Recently, many researchers have begun to try to improve the thermal conductivity of the interface by using heat treatment or designing double carbide interfaces. Although the current stage has provided limited help in improving the thermal conductivity of copper-based diamond composite materials, it is of significant importance for subsequent research.

5.3. Effect of Temperature, Pressure and Holding Time

In order to achieve copper-based diamond composites with high thermal conductivity it is important to select a suitable process temperature, because different temperatures can result in different states for copper matrix and diamond particles. Figure 12 depicts the microstructures of the fracture surfaces of copper-based diamond composites that were sintered at various temperatures. The localized interfacial debonding was found at a low temperature of 860 °C, while the interfacial adhesion was improved significantly when the temperature increased to 920 °C. However, when the temperature increased continuously to 940 °C, the interfacial bonding deteriorated again [55]. The reason is that at low sintering temperatures, the gaps between diamond particles are not well filled and the poor fluidity of copper can result in low thermal conductivity of copper-based diamond composites. At 920 °C, the fluidity of copper improves, leading to an increase in the density and thermal conductivity of copper-based diamond composites. At high temperatures, however, the following conditions may lead to a decrease in the thermal conductivity. Firstly, the diamond surface may undergo graphitization due to high temperature. Additionally, thermal stresses on diamond particles manifest as compressive stresses during melting, and tensile stresses during curing, causing the diamond particles to detach from the copper matrix easily.

Figure 12. Fracture surfaces of Cu/diamond composites sintered at: (**a**) 860 °C; (**b**) 920 °C; (**c**) 940 °C. Cu-Cr/diamond composites sintered at: (**d**) 860 °C; (**e**) 920 °C; (**f**) 940 °C. Copyright 2010, with permission from Elsevier [55].

Pressure is also a crucial factor in determining the thermal conductivity of copper-based diamond composites. As the sintering pressure increases, molten copper can penetrate the interstices of diamond particles, which enhances the densities and thermal conductivity of copper-based diamond composites. However, excessive pressure can decrease the density of copper-based diamond composites. Figure 13 shows that the pores in the copper-based diamond composites obtained at 550 MPa are more evident than those obtained at 525 MPa [56]. The possible reason for this phenomenon is that high pressure can lead to gas spillage, creating cavities in the composite and reducing the densities and thermal conductivity of copper-based diamond composites.

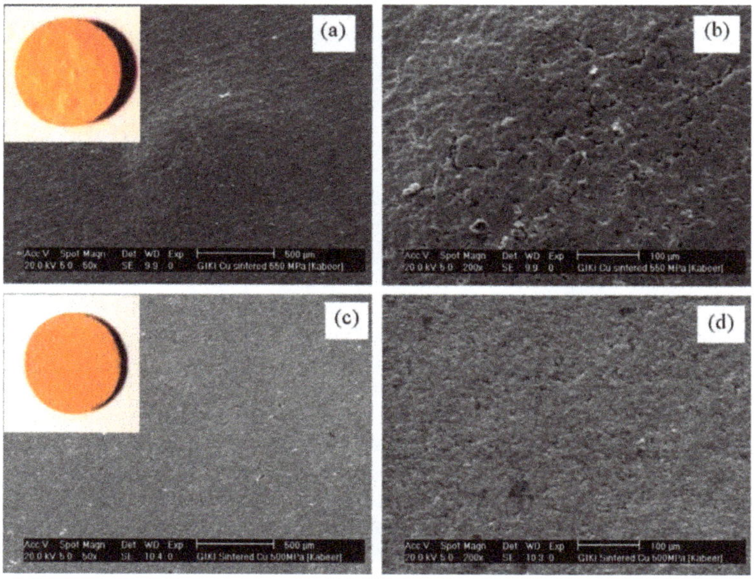

Figure 13. Microstructures of copper-based diamond composites compacted at: (**a**,**b**) 550 MPa; (**c**,**d**) 525 MPa. Inserting image in (**a**) shows the blisters on the surface and inserting image in (**c**) shows no blistering on the surface. Copyright 2014, with permission from Elsevier [56].

The holding time at high temperatures is also a crucial parameter. H. Li [57] used vacuum micro-plasma deposition to deposit a titanium coating on the diamond surface, which enhanced the interfacial binding between the diamond particles and copper. In this study, the effect of holding time on the density of the composite material was investigated. The results showed that the relative density of Ti-coated diamond/copper composite material increased at the beginning, and then decreased with an increase in holding time, reaching a maximum value of 98.72% at 40 min. The reason is that atomic diffusion and migration between Ti-coated diamond and copper substrate require sufficient holding time. Excessive holding time may result in grain growth, gaps, and cracks, leading to a decrease in the density of the composite material.

5.4. *Effects of Heat Treatment*

Interfacial modification is usually performed by coating the diamond surface with an element or by adding an alloying element to the copper matrix. The interface modification method enables formation of a carbonized layer between the diamond particles and copper. In general, controlling the thickness of the carbonized layer and increasing the density of the carbonized layer is the key to improving the thermal conductivity of the interface. In order to obtain a better quality interface, researchers have tried to improve the thermal conductivity of the diamond by means of heat treatment. J. Sang [58] used magnetron sputtering to deposit tungsten (W) coating on the surface of diamond particles. They prepared copper-based diamond composite materials with high thermal conductivity. The W-coated diamond particles were subjected to different annealing processes before the preparation of the composites by pressure-assisted infiltration. With an increase in annealing temperature, the metal W in the coating transforms into tungsten carbide (W_2C, WC). The thermal conductivity of the composites increased, and then decreased as the phase changed from metal W to tungsten carbide (W_2C, WC). The thermal conductivity increased from 656 W/(m·K) to 836 W/(m·K), and finally decreased to 770 W/(m·K). When the annealing temperature reached 1100 °C, the metal W disappeared. The coating on the diamond particles consisted of WC and W_2C phases. The combination of the diamond/WC and W_2C/Cu structures exhibited a higher interfacial thermal conductivity. The W_2C sublayer reduced the acoustic impedance mismatch between the Cu and the interfacial coating. When the annealing temperature reached 1200 °C, all the coatings on the diamond surface were transformed into WC, and the diamond was roughly graphitized, resulting, instead, in a decrease in thermal conductivity. Although heat treatment can change the interface structure between diamond and copper, not all heat treatment processes can increase the thermal conductivity of composite materials. L. Lei [59] prepared Cu-1.5 wt.% Ti/diamond (55 vol.%) composite material by hot forging. Through annealing, the thermal conductivity of the composite material was reduced. This is because annealing caused the TiC particles in the composite material to coarsen and agglomerate, as well as, it caused the interface layer of the composite material to crack and strip. In addition, it also caused a large amount of diamond to graphitize. Therefore, adopting a reasonable heat treatment process is one of the keys to improving the thermal conductivity of composite materials.

6. Forming Techniques

In the field of copper-based diamond research, interfacial modification of composite materials is one of the key components. Forming techniques are also a crucial factor for achieving copper-based diamond composites with high thermal conductivity. Currently, there are many mature forming techniques used to obtain copper-based diamond composites with high thermal conductivity. However, some common problems make them unsuitable for mass production. Researchers have been trying some new forming techniques. Although the thermal conductivity obtained through these techniques is not very high, they are still valuable for future research.

The high-temperature and high-pressure method (HTHP) involves filling uniformly mixed powder into graphite molds and sintering the copper/diamond composite materials at high temperature and high pressure. Generally speaking, when the diamond content is greater than 70 vol.%, it is not conducive to improving the thermal conductivity of the composite material. However, when using high-temperature and high-pressure sintering, copper easily fills the gaps between the diamond particles, resulting in high-density composite materials. At the same time, the high-temperature and high-pressure conditions make the preparation of copper/diamond faster. J. He [38] prepared copper-based composites using high-temperature and high-pressure methods. The composites were reinforced using about 90 volume percent diamond particles, and zirconium was added into the copper matrix. The Zr content varied between 0 and 2.0 wt%, to study the microstructure of the Cu-Zr/diamond composite material interface and thermal conductivity. The highest thermal conductivity of the composite material obtained was 677 W/(m·K). Although this is already quite a mature method, there are still many issues. The high-temperature and high-pressure conditions can lead to excessively high costs associated with preparing materials. In addition, due to the restrictions of graphite molds, the size of copper-based diamond composite materials is limited. The prepared sample is shown in Figure 14 [11].

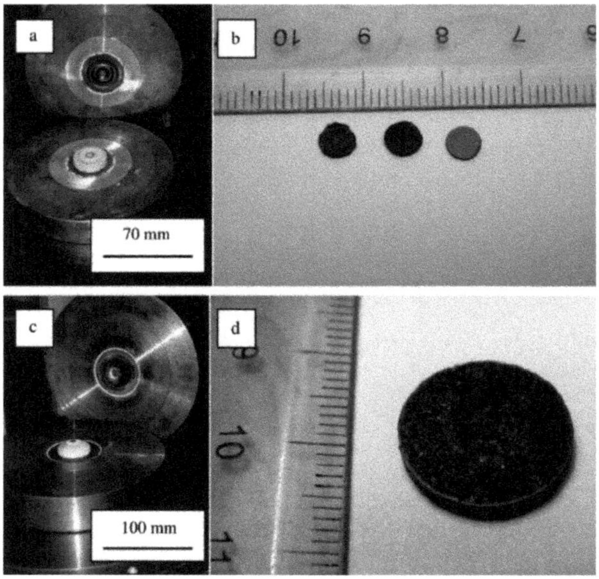

Figure 14. HPHT devices and composites produced with them: (**a**,**b**) toroid-type; (**c**,**d**) anvils-with-depression type apparatus. Copyright 2008, with permission from Elsevier [11].

Vacuum hot-pressing sintering (VHPS) is the process of mixing diamond and copper powder in a mold, and then preparing the composite material in a vacuum environment of a hot-pressing furnace. The vacuum hot-pressing sintering equipment consists of three parts: a vacuum system, a pressurization system, and a heating system. Figure 15 is a schematic diagram of the device [34]. Y. Pan [33] proposed a new method for preparing diamond/copper composite materials. Double-layer diamond particles were directly compressed under high pressure to prepare a preform, and then sintered in a vacuum device to densify it. The thermal conductivity of the obtained diamond (65 vol%)/Cu composite material was as high as 720 W/(m·K). H. Sun [60] used the vacuum hot-pressing sintering process to obtain dense diamond/Cu composite material blocks with different Cu volume fractions. The thermal conductivity of the obtained diamond/Cu composite material reached 638 W/(m·K), and the flexural strength reached 276 MPa. The mechanical properties and thermodynamic properties were excellent. The advantages of this method are that the heating rate is uniform and the heating rate is slow. It can effectively reduce the thermal stress generated in the sintering process of the composite material. However, at the same time, its preparation efficiency is correspondingly reduced. This method is also limited by the mold. The pressure cannot be too high, since the density of the composite material can be affected by the pressure.

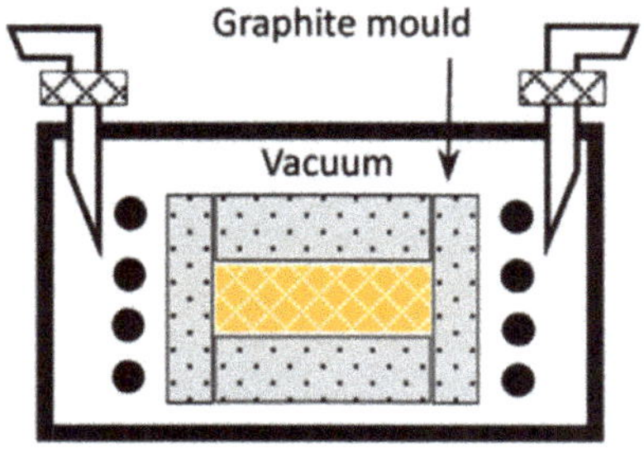

Figure 15. Schematic diagram of vacuum hot pressing sintering. Copyright 2018, with permission from Elsevier [34].

The spark plasma sintering (SPS) process involves placing uniformly mixed powders into a graphite mold and sintering them under the action of pulsed current and pressure. The sintering temperature of this method is usually between 800 and 970 °C, which is lower than the melting point of copper. Moreover, this method can rapidly produce high thermal conductivity composite material. It has a fast preparation speed and high efficiency. D. Zhao [36] used a mixture of diamond particles (with Cr/Cu coating) and Cu powder to produce a diamond/Cu composite material using spark plasma sintering. The time to complete the heating process was less than half an hour, and the thermal conductivity of the final sample reached 613 W/(m·K). Y. Pan [61] used a magnetron sputtering technique to coat diamond particles with a layer of zirconium. A diamond-copper composite material was prepared using spark plasma sintering. This method required heating to 900 °C and holding for 20 min, with a heating rate of 100 °C/min. The maximum thermal conductivity (TC) obtained was 609 W/(m·K), and the corresponding coefficient of thermal expansion (CTE) was as low as 6.75×10^{-6} K^{-1}. However, the size of the composite material was limited due to the restrictions of the graphite mold, and the applied pressure was also limited, resulting in low density of the composite material. Figure 16 shows a schematic diagram of the SPS device [62].

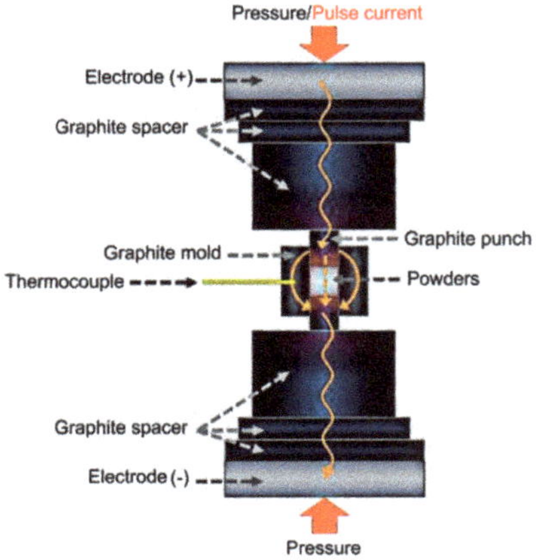

Figure 16. A schematic view of the SPS apparatus. Copyright 2017, with permission from Elsevier [62].

The molten infiltration method involves heating copper in a mixed powder to a molten state and gradually infiltrating it into the gaps of the diamond reinforcing particles. Depending on whether pressure is applied, it is divided into pressureless infiltration (PLI) and pressure infiltration (PI). Pressureless infiltration refers to molten copper permeating into the diamond reinforcing particles without the application of pressure. This method generally involves using a binder to prepare the reinforcement in a preform, heating the copper above its melting point, and then infiltrating it into the diamond reinforcing particles under the protection of a gas atmosphere. The advantage of this method is its simplicity and ease of operation. However, the use of a binder leads to a decrease in the thermal conductivity of the matrix and the resulting composite material has low density. S. Dai [63] prepared a tungsten, diamond, and copper composite material by pressureless infiltration. The composite material contained a large number of pores and defects. The effect of pressureless forming was poor, and compared to pressure infiltration, pressureless infiltration could not effectively fill the gaps between the diamond particles. Pressure infiltration involves infiltration of molten copper into diamond reinforcing particles under the action of pressure. In 2018, L. Wang [21] prepared a composite material of Cu/diamond (using Ti-coated diamond particles with coating thicknesses ranging from 65 nm to 850 nm) by using pressure infiltration. By adjusting the thickness of the interface TiC layer, the thermal conductivity of the Cu/diamond composite was improved. When the thickness of the Ti coating reached 220 nm, the maximum thermal conductivity value of the composite material was 811 W/(m·K). In 2019, L. Wang [29] prepared a Zr alloying-Cu matrix composite material reinforced with diamond particles using pressure infiltration, and the thermal conductivity of the composite material was 930 W/(m·K).

The above-mentioned methods are all mature, and the copper-based diamond composite materials produced have high thermal conductivity. However, it is difficult to produce complex-shaped composite materials using these methods. Recently, researchers have begun to study new methods to address production issues associated with copper-based diamond composite materials, such as additive manufacturing. L. Constantin [64] prepared a Cu/diamond composite with a thermal conductivity of 330 W/(m·K) using the laser-directed energy deposition method. The experimental results showed that the diamond did not graphitize, which demonstrated the feasibility of this method. The limitation of additive manufacturing is the slow production speed, high fixed price of solid materials, and limited material composition. In addition, this method is prone to defects, and the resulting thermal conductivity of the composite materials is not very high.

Typically, the methods for preparing copper-based diamond composite materials are time-consuming or energy-consuming. Hot forging is a feasible and economically effective method for preparing copper-based diamond composite materials, which has potential as a preparation method. The produced process has high mechanical and thermophysical properties, which has attracted the attention of many researchers in recent years. S. Q. Jia [65] successfully prepared Cu-Cr/55 vol% diamond composite by using hot forging powder preforms. As the Cr content increased from 1 wt% to 3 wt%, the relative density of the prepared composite increased. The thermal conductivity also increased accordingly. When the Cr content was 3 wt%, the thermal conductivity reached 487 W/(m·K). L. Lei [66] prepared Cu/55 vol% diamond (Ti) composite via hot forging. The thermal conductivity, CTE, and bending strength of the composite were 550 W/(m·K) (99% of the theoretical value), 7×10^{-6}/K, and 418 MPa, respectively. However, this method also has some issues. Copper-bassed diamond composite materials introduce significant internal stresses during hot forging, mainly limited between the interface and the copper matrix. The relatively fast cooling rate from forging temperature to room temperature can lead to interface detachment [67]. Suitable temperature and methods need to be found to control interface detachment between the interface and the matrix.

In addition to the methods mentioned above, many shaping methods have been used by researchers in recent years, such as the liquid–solid separation technique [68,69], friction stir processing [70], and ultrasonic consolidation [71]. Although the composite materials obtained through these methods are not the most outstanding, these attempts are of significant importance for future copper/diamond research.

7. Conclusions and Outlook

Due to the increasing integration of electronic devices, there is a growing demand for packaging materials with high thermal conductivity. However, conventional packaging materials fail to meet this high thermal conductivity requirement. To address this requirement, a copper-based diamond composite is proposed, where diamond particles (with a thermal conductivity of 2000 W/(m·K)) are compounded with copper matrix (with a thermal conductivity of 398 W/(m·K)). By enhancing the interface between copper and diamond particles, the thermal conductivity of copper-based diamond composites can be increased to over 900 W/(m·K). The exceptional thermal conductivity of copper-based diamond composites makes them ideal materials for use in military, aviation, and aerospace applications. Nevertheless, for mass production of copper-based diamond composites, there is a need to discover the suitable process parameters and preparation methods that can achieve high thermal conductivity while minimizing manufacturing costs. Ongoing studies on copper-based diamond composites have mainly focused on improving the interfaces of copper matrix and diamond particles, and optimizing the manufacturing processes with lower costs. Based on the conclusions of recent reviews [72,73] and our own perspective, the following recommendations can be made:

(1) There is a need to improve the existing models and to improve the preparation process of Cu/diamond composite materials. Based on our findings, there is still a significant gap between the theoretical models and the actual data. One method is to consider more factors to make the theoretical thermal conductivity closer to the actual thermal conductivity; however, this method would make the predictive model more complicated. Researchers can further improve the preparation process of Cu/diamond composite materials so that the thermal conductivity of the prepared materials is closer to the theoretical value. This would make the theoretical model more practical in guiding actual production.

(2) Studies on forming techniques that focus on reducing power consumption, improving efficiency, and increasing thermal conductivity lay a foundation for better practical applications in production. The existing shaping methods are not very efficient. High-power equipment is required to achieve ultra-high thermal conductivity. In recent years, although shaping methods have been working on solving power consumption and efficiency issues, the obtained composite materials do not have outstanding thermal conductivity. To make copper-based diamond composite materials more widely used in daily life, we need to pay more attention to the development of new forming techniques.

(3) Future studies on the copper-based diamond composites should focus on both the thermal and the mechanical properties. The existing studies have focused primarily on how to improve thermal conductivity, and many composite materials with high thermal conductivity have been developed. However, there are still many issues to be addressed before high thermal conductivity composite materials can be applied. We need to consider whether the strength of the composite material is sufficient. Similarly, we need to consider the processability of the composite material. Only when high thermal conductivity and high mechanical performance are simultaneously achieved, can copper-based diamond composite materials be more effectively applied in the field of electronic packaging.

Author Contributions: Conceptualization, H.C.; writing—original draft preparation, K.C.; writing—review and editing, R.Z., Y.K. and H.C.; supervision, X.L. and H.C.; funding acquisition, H.C. All authors have read and agreed to the published version of the manuscript.

Funding: The authors are grateful for the financial support of the Open Project Program of State Key Laboratory of Advanced Welding and Joining (No. AWJ-23Z04), the Guangdong Basic and Applied

Basic Research Foundation (No. 2021A1515012411), and the Characteristic Innovation Project of Guangdong Educational Department (No. 2022KTSCX217).

Data Availability Statement: The data presented in this study are available on request from the corresponding author.

Conflicts of Interest: The authors declare no conflict of interest.

References

1. Sarvar, F.; Whalley, D.C.; Conway, P.P. Thermal interface materials-A review of the state of the art. In Proceedings of the 2006 1st Electronic System Integration Technology Conference, Dresden, Germany, 5–7 September 2006; IEEE: New York, NY, USA, 2006; Volume 2, pp. 1292–1302.
2. Moore, A.L.; Shi, L. Emerging challenges and materials for thermal management of electronics. *Mater. Today* **2014**, *17*, 163–174. [CrossRef]
3. Yamamoto, Y.; Imai, T.; Tanabe, K.; Tsuno, T.; Kumazawa, Y.; Fujimori, N. The measurement of thermal properties of diamond. *Diam. Relat. Mater.* **1997**, *6*, 1057–1061. [CrossRef]
4. Ma, S.; Zhao, N.; Shi, C.; Liu, E.; He, C.; He, F.; Ma, L. Mo2C coating on diamond: Different effects on thermal conductivity of diamond/Al and diamond/Cu composites. *Appl. Surf. Sci.* **2017**, *402*, 372–383. [CrossRef]
5. Xue, C.; Yu, J.K.; Zhu, X.M. Thermal properties of diamond/SiC/Al composites with high volume fractions. *Mater. Des.* **2011**, *32*, 4225–4229. [CrossRef]
6. Liang, X.; Jia, C.; Chu, K.; Chen, H.; Nie, J.; Gao, W. Thermal conductivity and microstructure of Al/diamond composites with Ti-coated diamond particles consolidated by spark plasma sintering. *J. Compos. Mater.* **2012**, *46*, 1127–1136. [CrossRef]
7. Wu, J.; Zhang, H.; Zhang, Y.; Li, J.; Wang, X. Effect of copper content on the thermal conductivity and thermal expansion of Al–Cu/diamond composites. *Mater. Des.* **2012**, *39*, 87–92. [CrossRef]
8. Ren, S.; Shen, X.; Guo, C.; Liu, N.; Zang, J.; He, X.; Qu, X. Effect of coating on the microstructure and thermal conductivities of diamond–Cu composites prepared by powder metallurgy. *Compos. Sci. Technol.* **2011**, *71*, 1550–1555. [CrossRef]
9. Zhang, Y.J.; Dong, Y.H.; Zhang, R.Q.; Ren, X.T.; Chu, A.M.; Chen, D.Z.; Chen, Q.J.; Ye, Z.G. Numerical simulation of thermal conductivity of diamond/copper composites. *Trans. Mater. Heat Treat.* **2018**, *39*, 110.
10. Schubert, T.; Trindade, B.; Weißgärber, T.; Kieback, B. Interfacial design of Cu-based composites prepared by powder metallurgy for heat sink applications. *Mater. Sci. Eng. A* **2008**, *475*, 39–44. [CrossRef]
11. Ekimov, E.A.; Suetin, N.V.; Popovich, A.F.; Ralchenko, V.G. Thermal conductivity of diamond composites sintered under high pressures. *Diam. Relat. Mater.* **2008**, *17*, 838–843. [CrossRef]
12. Zhang, C.; Cai, Z.; Tang, Y.; Wang, R.; Peng, C.; Feng, Y. Microstructure and thermal behavior of diamond/Cu composites: Effects of surface modification. *Diam. Relat. Mater.* **2018**, *86*, 98–108. [CrossRef]
13. Zhang, X.Y.; Xu, M.; Cao, S.Z.; Chen, W.B.; Yang, W.Y.; Yang, Q.Y. Enhanced thermal conductivity of diamond/copper composite fabricated through doping with rare-earth oxide Sc_2O_3. *Diam. Relat. Mater.* **2020**, *104*, 107755. [CrossRef]
14. Chao, Z.; Wang, J. Enhanced mechanical properties in diamond/Cu composites with chromium carbide coating for structural applications. *Mater. Sci. Eng. A* **2013**, *588*, 221–227.
15. Eucken, A. Heat transfer in ceramic refractory materials: Calculation from thermal conductivities of constituents. *Fortchg. Gebiete Ingenieurw. B3 Forschungsheft* **1932**, *16*, 353–360.
16. Hasselman, D.P.H.; Donaldson, K.Y.; Liu, J.; Gauckler, L.J.; Ownby, P.D. Thermal Conductivity of a Particulate-Diamond-Reinforced Cordierite Matrix Composite. *J. Am. Ceram. Soc.* **1994**, *77*, 1757–1760. [CrossRef]
17. Tavangar, R.; Molina, J.M.; Weber, L. Assessing predictive schemes for thermal conductivity against diamond-reinforced silver matrix composites at intermediate phase contrast. *Scr. Mater.* **2007**, *56*, 357–360. [CrossRef]
18. Zhang, Y.; Zhang, H.L.; Wu, J.H.; Wang, X.T. Enhanced thermal conductivity in copper matrix composites reinforced with titanium-coated diamond particles. *Scr. Mater.* **2011**, *65*, 1097–1100. [CrossRef]
19. Che, Q.L.; Zhang, J.J.; Chen, X.K.; Ji, Y.Q.; Li, Y.W.; Wang, L.X.; Cao, S.Z.; Guo, L.; Wang, Z.; Wang, S.W.; et al. Spark plasma sintering of titanium-coated diamond and copper–titanium powder to enhance thermal conductivity of diamond/copper composites. *Mater. Sci. Semicond. Process.* **2015**, *33*, 67–75. [CrossRef]
20. Zhang, Y.; Bai, G.; Liu, X.; Dai, J.; Wang, X.; Zhang, H. Reinforcement size effect on thermal conductivity in Cu-B/diamond composite. *J. Mater. Sci. Technol.* **2021**, *91*, 1–4. [CrossRef]
21. Wang, L.; Li, J.; Catalano, M.; Bai, G.; Li, N.; Dai, J.; Wanga, X.; Zhanga, H.; Kim, M.J. Enhanced thermal conductivity in Cu/diamond composites by tailoring the thickness of interfacial TiC layer. *Compos. Part A Appl. Sci. Manuf.* **2018**, *113*, 76–82. [CrossRef]
22. Rape, A.; Gott, K.; Kulkarni, A.; Singh, J. Simulation of matrix conductivity in copper–diamond composites sintered by field assisted sintering technology. *Comput. Mater. Sci.* **2015**, *110*, 29–33. [CrossRef]
23. Every, A.G.; Tzou, Y.; Hasselman, D.P.H.; Raj, R. The Effect of Microstructure on the Thermal Conductivity of Particulate ZnS/Diamond Composites. *Acta Metall. Mater.* **1992**, *40*, 123–129. [CrossRef]
24. Swartz, E.T.; Pohl, R.O. Thermal boundary resistance. *Rev. Mod. Phys.* **1989**, *61*, 605. [CrossRef]

25. Zhang, Y.; Wang, Z.; Li, N.; Che, Z.; Liu, X.; Chang, G.; Hao, J.; Dai, J.; Wang, X.; Sun, F.; et al. Interfacial Thermal Conductance between Cu and Diamond with Interconnected W− W2C Interlayer. *ACS Appl. Mater. Interfaces* **2022**, *14*, 35215–35228. [CrossRef] [PubMed]
26. Kidalov, S.V.; Shakhov, F.M.; Vul, A.Y. Thermal conductivity of nanocomposites based on diamonds and nanodiamonds. *Diam. Relat. Mater.* **2007**, *16*, 2063–2066. [CrossRef]
27. Hanada, K.; Matsuzaki, K.; Sano, T. Thermal properties of diamond particle-dispersed Cu composites. *J. Mater. Process. Technol.* **2004**, *153*, 514–518. [CrossRef]
28. Wu, Y.; Luo, J.; Wang, Y.; Wang, G.; Wang, H.; Yang, Z.; Ding, G. Critical effect and enhanced thermal conductivity of Cu-diamond composites reinforced with various diamond prepared by composite electroplating. *Ceram. Int.* **2019**, *45*, 13225–13234. [CrossRef]
29. Wang, L.; Li, J.; Bai, G.; Li, N.; Wang, X.; Zhang, H.; Wang, J.; Kim, M.J. Interfacial structure evolution and thermal conductivity of Cu-Zr/diamond composites prepared by gas pressure infiltration. *J. Alloy. Compd.* **2019**, *781*, 800–809. [CrossRef]
30. Zhang, Y.; Bai, G.; Zhu, X.; Dai, J.; Wang, X.; Wang, J.; Kim, M.J.; Zhang, H. Manipulating in-situ discrete carbide interlayer to achieve high thermal conductivity in Cu-B/diamond composite. *Mater. Today Commun.* **2023**, *34*, 105357. [CrossRef]
31. Bai, G.; Li, N.; Wang, X.; Wang, J.; Kim, M.J.; Zhang, H. High thermal conductivity of Cu-B/diamond composites prepared by gas pressure infiltration. *J. Alloy. Compd.* **2018**, *735*, 1648–1653. [CrossRef]
32. Wang, L.; Li, J.; Che, Z.; Wang, X.; Zhang, H.; Wang, J.; Kim, M.J. Combining Cr pre-coating and Cr alloying to improve the thermal conductivity of diamond particles reinforced Cu matrix composites. *J. Alloy. Compd.* **2018**, *749*, 1098–1105. [CrossRef]
33. Pan, Y.; He, X.; Ren, S.; Wu, M.; Qu, X. High thermal conductivity of diamond/copper composites produced with Cu–ZrC double-layer coated diamond particles. *J. Mater. Sci.* **2018**, *53*, 8978–8988. [CrossRef]
34. Pan, Y.; He, X.; Ren, S.; Wu, M.; Qu, X. Optimized thermal conductivity of diamond/Cu composite prepared with tungsten-copper-coated diamond particles by vacuum sintering technique. *Vacuum* **2018**, *153*, 74–81. [CrossRef]
35. Xie, Z.; Guo, H.; Zhang, X.; Huang, S.; Xie, H.; Mi, X. Tailoring the thermal and mechanical properties of diamond/Cu composites by interface regulation of Cr alloying. *Diam. Relat. Mater.* **2021**, *114*, 108309. [CrossRef]
36. Zhao, D.; Zha, S.; Liu, D. Influence of sputtering and electroless plating of Cr/Cu dual-layer structure on thermal conductivity of diamond/copper composites. *Diam. Relat. Mater.* **2021**, *115*, 108296. [CrossRef]
37. Tao, J.M.; Zhu, X.K.; Tian, W.W.; Peng, Y.A.; Hao, Y.A. Properties and microstructure of Cu/diamond composites prepared by spark plasma sintering method. *Trans. Nonferrous Met. Soc. China* **2014**, *24*, 3210–3214. [CrossRef]
38. He, J.; Wang, X.; Zhang, Y.; Zhao, Y.; Zhang, H. Thermal conductivity of Cu–Zr/diamond composites produced by high temperature–high pressure method. *Compos. Part B Eng.* **2015**, *68*, 22–26. [CrossRef]
39. Xie, Z.; Guo, H.; Zhang, Z.; Zhang, X. Thermal expansion behavior and dimensional stability of Diamond/Cu composites with different diamond content. *J. Alloy. Compd.* **2019**, *797*, 122–130. [CrossRef]
40. Jia, J.; Bai, S.; Xiong, D.; Wang, J.; Chang, J. Effect of tungsten based coating characteristics on microstructure and thermal conductivity of diamond/Cu composites prepared by pressueless infiltration. *Ceram. Int.* **2019**, *45*, 10810–10818. [CrossRef]
41. Chang, G.; Sun, F.; Duan, J.; Che, Z.; Wang, X.; Wang, J.; Kim, M.J.; Zhang, H. Effect of Ti interlayer on interfacial thermal conductance between Cu and diamond. *Acta Mater.* **2018**, *160*, 235–246. [CrossRef]
42. Yang, L.; Sun, L.; Bai, W.; Li, L. Thermal conductivity of Cu-Ti/diamond composites via spark plasma sintering. *Diam. Relat. Mater.* **2019**, *94*, 37–42. [CrossRef]
43. Chu, K.; Jia, C.; Guo, H.; Li, W. On the thermal conductivity of Cu–Zr/diamond composites. *Mater. Des.* **2013**, *45*, 36–42. [CrossRef]
44. Sang, J.; Yang, W.; Zhu, J.; Fu, L.; Li, D.; Zhou, L. Regulating interface adhesion and enhancing thermal conductivity of diamond/copper composites by ion beam bombardment and following surface metallization pretreatment. *J. Alloy. Compd.* **2018**, *740*, 1060–1066. [CrossRef]
45. Wu, X.; Li, L.; Zhang, W.; Song, M.; Yang, W.; Peng, K. Effect of surface roughening on the interfacial thermal conductance of diamond/copper composites. *Diam. Relat. Mater.* **2019**, *98*, 107467. [CrossRef]
46. Wu, X.; Wan, D.; Zhang, W.; Song, M.; Peng, K. Constructing efficient heat transfer channels at the interface of Diamond/Cu composites. *Compos. Interfaces* **2021**, *28*, 625–635. [CrossRef]
47. Zhang, C.; Wang, R.; Cai, Z.; Peng, C.; Feng, Y.; Zhang, L. Effects of dual-layer coatings on microstructure and thermal conductivity of diamond/Cu composites prepared by vacuum hot pressing. *Surf. Coat. Technol.* **2015**, *277*, 299–307. [CrossRef]
48. Li, H.; Xie, Y.; Zhang, L.; Wang, H. Thermal properties of diamond/Cu composites enhanced by TiC plating with molten salts containing fluoride and electroless-plated Cu on diamond particles. *Diam. Relat. Mater.* **2022**, *129*, 109337. [CrossRef]
49. Sang, J.; Yuan, Y.; Yang, W.; Zhu, J.; Fu, L.; Li, D.; Zhou, L. Exploring the underlying causes of optimizing thermal conductivity of copper/diamond composites by interface thickness. *J. Alloy. Compd.* **2022**, *891*, 161777. [CrossRef]
50. Jia, J.; Bai, S.; Xiong, D.; Xiao, J.; Yan, T. Enhanced thermal conductivity in diamond/copper composites with tungsten coatings on diamond particles prepared by magnetron sputtering method. *Mater. Chem. Phys.* **2020**, *252*, 123422. [CrossRef]
51. Kang, Q.; He, X.; Ren, S.; Liu, T.; Liu, Q.; Wu, M.; Qu, X. Microstructure and thermal properties of copper–diamond composites with tungsten carbide coating on diamond particles. *Mater. Charact.* **2015**, *105*, 18–23. [CrossRef]
52. Kang, Q.; He, X.; Ren, S.; Zhang, L.; Wu, M.; Guo, C.; Cui, W.; Qu, X. Preparation of copper–diamond composites with chromium carbide coatings on diamond particles for heat sink applications. *Appl. Therm. Eng.* **2013**, *60*, 423–429. [CrossRef]

53. Fu, X.; Jiang, J.; Jiang, X. Research Progress in Interfacial Characteristics and Strengthening Mechanisms of Rare Earth Metal Oxide-Reinforced Copper Matrix Composites. *Materials* **2022**, *15*, 5350. [CrossRef]
54. Kang, Q.; Wu, X. Preparation, mechanical properties and thermo physical properties of copper/diamond composites with WC coating diamond particles by vacuum-pressure infiltration. *J. Alloy. Compd.* **2019**, *789*, 310–316.
55. Chu, K.; Liu, Z.; Jia, C.; Chen, H.; Liang, X.; Gao, W.; Tian, W.; Guo, H. Thermal conductivity of SPS consolidated Cu/diamond composites with Cr-coated diamond particles. *J. Alloy. Compd.* **2010**, *490*, 453–458. [CrossRef]
56. Raza, K.; Khalid, F.A. Optimization of sintering parameters for diamond–copper composites in conventional sintering and their thermal conductivity. *J. Alloy. Compd.* **2014**, *615*, 111–118. [CrossRef]
57. Li, H.; Wang, C.; Wu, L.; Chen, M.; Wu, C.; Wang, N.; Li, Z.; Tang, L.; Pang, Q. Optimization of process parameters, microstructure, and thermal conductivity properties of Ti-coated diamond/copper composites prepared by spark plasma sintering. *J. Mater. Sci. Mater. Electron.* **2021**, *32*, 9115–9125. [CrossRef]
58. Sang, J.; Zhou, L.; Yang, W.; Zhu, J.; Fu, L.; Li, D. Enhanced thermal conductivity of copper/diamond composites by fine-regulating microstructure of interfacial tungsten buffer layer. *J. Alloy. Compd.* **2021**, *856*, 157440. [CrossRef]
59. Lei, L.; Su, Y.; Bolzoni, L.; Yang, F. Evaluation on the interface characteristics, thermal conductivity, and annealing effect of a hot-forged Cu-Ti/diamond composite. *J. Mater. Sci. Technol.* **2020**, *49*, 7–14. [CrossRef]
60. Sun, H.; Guo, L.; Deng, N.; Li, X.; Li, J.; He, G.; Li, J. Elaborating highly thermal-conductive diamond/Cu composites by sintering intermittently electroplated core-shell powders. *J. Alloy. Compd.* **2019**, *810*, 151907. [CrossRef]
61. Pan, Y.; He, X.; Ren, S.; Wu, M.; Qu, X. Improvement of ZrC/Zr coating on the interface combination and physical properties of diamond-copper composites fabricated by spark plasma sintering. *Materials* **2019**, *12*, 475. [CrossRef]
62. Azarniya, A.; Azarniya, A.; Sovizi, S.; Hosseini, H.R.M.; Varol, T.; Kawasaki, A.; Ramakrishna, S. Physicomechanical properties of spark plasma sintered carbon nanotube-reinforced metal matrix nanocomposites. *Prog. Mater. Sci.* **2017**, *90*, 276–324. [CrossRef]
63. Dai, S.G.; LI, J.W.; Wang, C.J. Preparation and thermal conductivity of tungsten coated diamond/copper composites. *Trans. Nonferrous Met. Soc. China* **2022**, *32*, 2979–2992. [CrossRef]
64. Constantin, L.; Fan, L.; Pontoreau, M.; Wang, F.; Cui, B.; Battaglia, J.L.; Silvain, J.F.; Lu, Y.F. Additive manufacturing of copper/diamond composites for thermal management applications. *Manuf. Lett.* **2020**, *24*, 61–66. [CrossRef]
65. Jia, S.Q.; Bolzoni, L.; Li, T.; Yang, F. Unveiling the interface characteristics and their influence on the heat transfer behavior of hot-forged Cu–Cr/Diamond composites. *Carbon* **2021**, *172*, 390–401. [CrossRef]
66. Lei, L.; Bolzoni, L.; Yang, F. High thermal conductivity and strong interface bonding of a hot-forged Cu/Ti-coated-diamond composite. *Carbon* **2020**, *168*, 553–563. [CrossRef]
67. Lei, L.; Bolzoni, L.; Yang, F. The Impact of Interface Characteristics on Mechanical Performance of a Hot-Forged Cu/Ti-Coated-Diamond Composite. *Mater. Sci. Forum* **2021**, *1016*, 1682–1689. [CrossRef]
68. Li, Y.; Zhou, H.; Wu, C.; Yin, Z.; Liu, C.; Liu, J.; Shi, Z. Interfacial Characterization and Thermal Conductivity of Diamond/Cu Composites Prepared by Liquid-Solid Separation Technique. *Nanomaterials* **2023**, *13*, 878. [CrossRef]
69. Zhou, H.; Ran, M.; Li, Y.; Yin, Z.; Tang, Y.; Zhang, W.; Zheng, W.; Liu, J. Improvement of thermal conductivity of diamond/Al composites by optimization of liquid-solid separation process. *J. Mater. Process. Technol.* **2021**, *297*, 117267. [CrossRef]
70. Huda, N.; Bisht, A.; Moreau, E.; Corbin, S.; Rabkin, E.; Gerlich, A.P. Fabrication of copper–diamond composite by friction stir processing. *J. Mater. Sci.* **2023**, *58*, 4184–4198. [CrossRef]
71. Wang, B.; Zhang, H.; Yu, G.; Du, R.; He, P. A novel ultrasonic consolidation method for rapid preparation of diamond/Cu composites. *Mater. Lett.* **2022**, *323*, 132498. [CrossRef]
72. Dai, S.; Li, J.; Lu, N. Research progress of diamond/copper composites with high thermal conductivity. *Diam. Relat. Mater.* **2020**, *108*, 107993. [CrossRef]
73. Jia, S.Q.; Yang, F. High thermal conductive copper/diamond composites: State of the art. *J. Mater. Sci.* **2021**, *56*, 2241–2274. [CrossRef]

MDPI AG
Grosspeteranlage 5
4052 Basel
Switzerland
Tel.: +41 61 683 77 34

Crystals Editorial Office
E-mail: crystals@mdpi.com
www.mdpi.com/journal/crystals

www.ingramcontent.com/pod-product-compliance
Lightning Source LLC
LaVergne TN
LVHW070446061126
851858LV00011B/169
* 9 7 8 3 7 2 5 8 7 6 9 9 0 *